PRINCIPLES OF PRACTICE

VOLUME VI: AI AND AUTOMATION OPERATIONS

IT WAS WORKING, AND NOBODY COULD TELL

JT Taylor, MS, MBA

National Board of Operations Professionals

Purchase of this book is not required in order to sit any National Board of Operations Professionals assessment and confers no advantage. The assessment is built from the published standard, not from this book.

NATIONAL BOARD OF OPERATIONS PROFESSIONALS

Principles of Practice, Volume VI: AI and Automation Operations

Published by the National Board of Operations Professionals, 16220 N. Scottsdale Rd, Suite 300, Scottsdale, AZ 85254, United States.

nationalboardofoperations.org

ISBN (paperback): 979-8-952466-05-0

First edition, 2026. Revision 1.0.

This volume is revised on a published cycle. Corrections and citation promotions are issued as versioned revisions of this edition rather than as new editions, and the revision number above identifies which text you are holding.

Purchase of this book is not required in order to sit any National Board of Operations Professionals assessment and confers no advantage. The assessment is built from the published standard, not from this book.

Set in DejaVu Serif and DejaVu Sans Condensed. Illustrations by the author.

Printed in the United States of America.

INTRODUCTION

ONE MONTH

A twenty agent real estate brokerage hired us in 2024 to automate part of their nurturing sequence. Email. The most ordinary automation there is.

It went in. It ran. Every message was delivered, correctly formatted, on schedule, to the right person, with the right merge fields populated from the right records. By every measure the system was capable of producing, it was working perfectly.

One month later the monthly analytics came back and open rates were down. Click through was down. Replies had almost stopped.

Nothing had broken. There was no error log to read, no failed job, no bounced batch, no alert. **The system could not tell us it had failed, because by its own definition it had not.** It had done exactly what it was built to do.

What it had not done was carry what the realtor knew. The buyer's wife. Their kid. The brother who came on the second tour and asked about the school district. The realtors had been putting that into their emails for years, and they had never mentioned it during discovery, because nobody had thought to describe it as part of the process. It was just what you say to someone you met on a Saturday.

The recipients could feel the difference immediately. They did not complain. They just stopped replying.

THE PART THAT SHOULD WORRY YOU

Here is what I want you to notice about that story, because it is not the part most people notice.

We found it in a month, and that was luck. We found it because a marketing sequence has an open rate. Somebody was already counting. If that same failure had happened inside a process with no natural metric attached to it, it would still be running.

And I want you to notice something harder. **The automation was correct.** If you had audited it against the specification, it would have passed. If you had asked the vendor whether it was performing, they would have said yes and shown you the delivery rate. The only check that would have caught it was a check nobody in that building was in a position to run: is this email carrying the thing that makes it work?

To run that check, you have to know what makes it work. The realtors knew. Nobody asked them in a way that got the answer out, including me.

WHAT THIS BOOK IS ABOUT

This is a book about the check.

Every book about automation you have read is about the task. What to automate, how to pick it, which tools, which order, what it costs, what it returns. This book covers that, because you cannot skip it. But it is not the argument.

The argument is that automation almost never fails at the task. It fails at the check.

The work gets done. It gets done on time, in volume, in the right format. And then it is wrong in some way that the system is structurally incapable of noticing, and it stays wrong for as long as it takes a human being to notice, and how long that takes depends entirely on whether anyone left in the building could have done the work themselves.

THE CLAIM

> ***You can only safely automate work that somebody still in the building could do themselves.***

Not because they should have to. Because the moment nobody can do it, nobody can tell whether it was done.

I want to be exact about what I am and am not saying, because this claim gets misread in a predictable direction.

I am not saying automate less. I automate constantly, for businesses making a thousand dollars a month and for businesses making a million a month and for companies you have heard of. I am not nostalgic about manual work and I do not think there is virtue in doing by hand what a machine does better.

I am saying that the capacity to check is a real asset with a real cost, that it decays, and that almost nobody accounts for it. It does not appear on any balance sheet. It is not in the business case. And it is the only thing standing between you and an automation that has been quietly wrong since March.

WHAT THIS BOOK WILL NOT DO

It will not tell you AI is a bubble, and it will not tell you AI changes everything. Both are positions people take instead of doing the work. I have no forecast for you.

It will not give you a failure rate for AI projects. There are four widely quoted ones and I took all four apart in Volume V. There is a fifth about jobs and automation that I take apart in chapter fifteen. **None of the five survives contact with its own source document**, and a book that opens by quoting one has already told you it did not check.

It will not name a single product. Not the model I use, not the model you use, not the ones in the studies except where a citation would be uncheckable without it. Products change every few months. What this book describes has been true since 1983 and I will show you the paper.

It will not resolve the question of what automation does to employment. I have run automations that cost people their roles. I will tell you about that in chapter fifteen and I am not going to pretend I have made peace with it.

WHAT YOU WILL GET

An instrument. One test, four conditions, applied to one task at a time, that tells you whether that task is safe to automate and whether it is still safe two years later.

Three of the four conditions are mine, from twenty years of doing this. **The fourth is not mine.** It comes from a paper published in a control engineering journal in 1983, it is the most important condition of the four, and I had never heard of it until this book was researched. I will tell you exactly when I learned it, because a book that pretends its author knew everything in advance is not a book about operations. It is an advertisement.

ONE PROMISE

By the end of chapter eight you will understand why the people best equipped to check an automation are also the people most confident that it is working, and why those two facts are the same fact.

That is the whole problem. Everything else is implementation.

THE AUTOMATION OPERATOR'S DIAGNOSTIC

Take one task. Not a role, not a department, not "our customer service." **One task**, the smallest complete unit of work you could hand to someone.

It can be one you already automated or one you are about to. Answer honestly. Nobody is watching.

PART ONE. THE TASK

1. Can you write down every step of this task, in order, without asking anyone?
2. Can you write down what the task does when the input is wrong, missing, or strange?
3. In the last ten times this task ran, how many needed a human to change something about how it was done?
4. If you handed this task to a competent new hire with the written steps and nothing else, would they produce the same output you would?

If you answered no to question one, stop. You are not automating a task, you are automating a guess.

PART TWO. THE STANDARD

1. Is there a written description of what a correct output looks like?
2. Is there a written description of what a **wrong** output looks like, that is more specific than "not correct"?
3. If the output were subtly wrong, in a way that still looked complete and professional, what would tip you off?
4. Where is that written down?

Question seven is the one that matters. If your answer is a feeling, you do not have a standard, you have a habit.

PART THREE. THE CHECKER

1. Name the person who checks this. An actual name.
2. Could that person do this task themselves, to the standard, today, without the automation?
3. When did they last do it themselves?
4. If they left, who would be able to answer question ten?

If nobody can be named at question nine, the task is unchecked and you did not know that until just now.

PART FOUR. THE DECAY

1. If this task stays automated for three years, will the person named at question nine still be able to do it?
2. What in your operation makes that true, other than hoping?
3. If the automation produced a wrong output tomorrow, how long would it take you to find out, and what would tell you?

Question thirteen is the one nearly everyone fails, and failing it does not mean stop. It means the check has an expiry date and you did not know it had one.

PART I

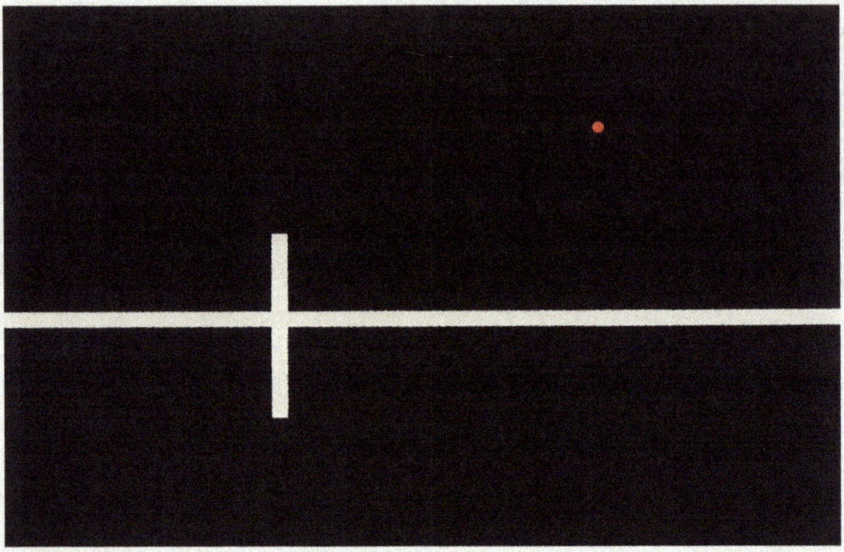

THE TASK

1. THE SEQUENCE THAT WORKED AND FAILED AT THE SAME TIME

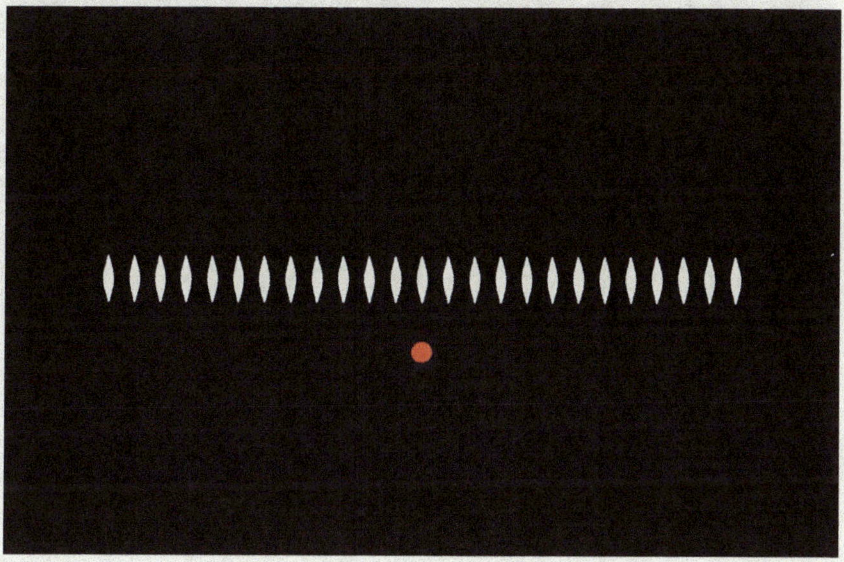

WHY THIS CHAPTER MATTERS

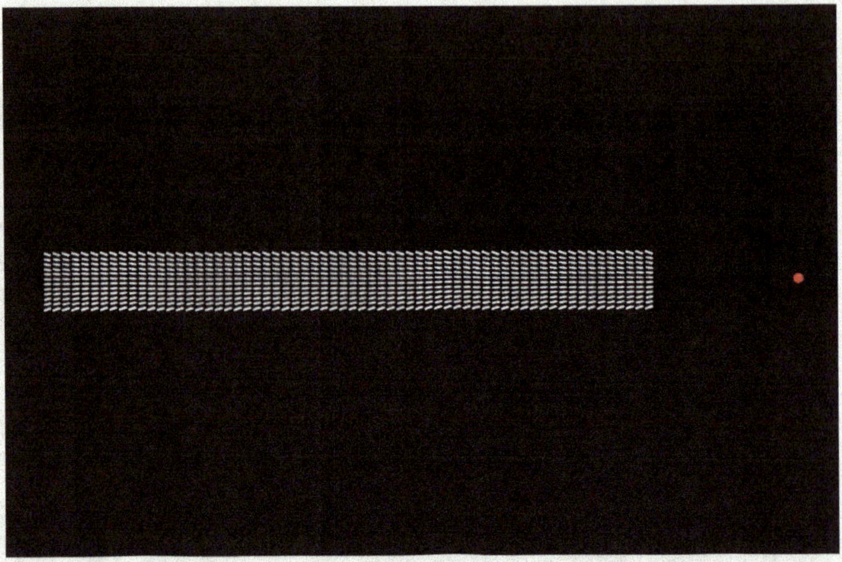

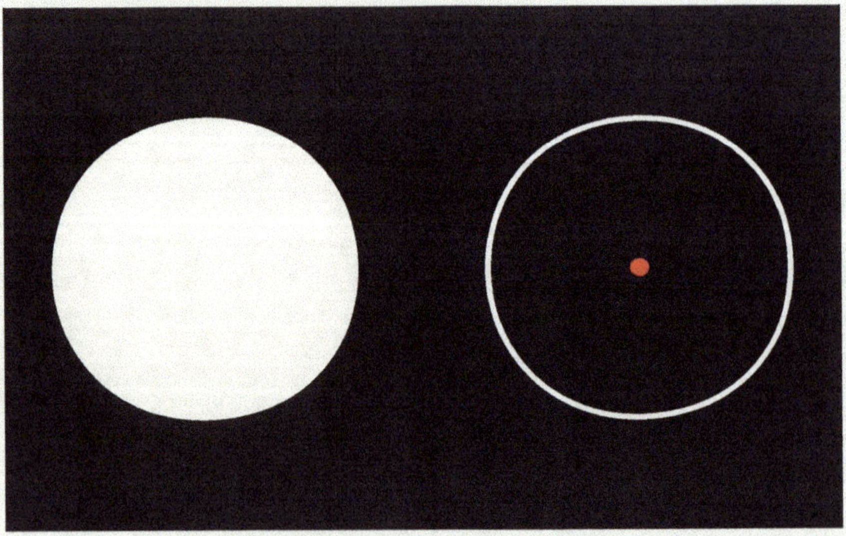

Because the failure in this chapter would have passed every audit you know how to run, and it cost a client real money for a month before anything surfaced. **A correct output and a right outcome are two different things, and most operational checking cannot tell them apart.**

WHAT WE WERE ASKED TO DO

The brokerage had twenty agents. Not a small business, not a large one. Big enough that the founder no longer knew what every agent did on a Tuesday, small enough that he thought he did.

The ask was straightforward. Buyers came through open houses and tours, agents followed up by email, and follow up was inconsistent. Some agents were disciplined and some were not. Deals were being lost in the gap between a tour on Saturday and a call on Thursday.

That is a textbook automation case, and I want to be clear that I would take it again. Sequential, high volume, templated, time sensitive, and demonstrably failing when done by hand. If you brought me that brief tomorrow I would say yes.

WHAT WE BUILT

We built what you would build. A sequence triggered by the tour record, branching on property type and price band, with merge fields for name, property, and the specifics of what they had seen.

We tested it. It fired correctly on every branch. The merge fields populated. The timing was right. We shipped it.

It ran exactly as specified for four weeks.

WHAT THE NUMBERS DID

The brokerage ran analytics monthly, which is how we found out at all.

Open rate fell. Click through fell harder. Replies, which is the only one that mattered, nearly stopped.

No part of the system reported a problem. Delivery was at a hundred percent. There were no bounces, no failed jobs, no exceptions in the log. If you had asked the platform whether the automation was healthy it would have said yes, and it would have been telling the truth about everything it could see.

WHAT WAS ACTUALLY MISSING

I sat down with three of the agents and read their old sent folders.

The emails they had been writing were not follow up emails. They were letters from someone who had spent ninety minutes with a family on a Saturday. They mentioned the wife by name. They mentioned that the kid had asked about the trampoline in the back garden. They mentioned that the brother in law had come to the second viewing and had strong opinions about the boiler.

None of that was in the process document, because none of it had ever been written down. When we asked how they followed up, they described the sequence. They did not describe the content, because the content was not a process to them. **It was just what you say to people you have met.**

NOBODY LIED TO US

This is the part I need you to sit with, because it is where the standard advice fails.

The client did not withhold anything. When we asked whether the follow up process was repeatable, they said yes, and they believed it, and by the definition in their heads they were right. There was a sequence. It had steps. The steps were the same every time.

They answered the question I asked. The question I asked was not the question I needed answered, and I did not know that until the open rate fell.

I have been doing this for twenty years. I asked good questions. I got true answers. And the automation was still wrong, because the thing that made the old process work had never been visible to anyone as a step, including the people doing it.

THE CHECK THAT DID NOT EXIST

Here is the whole book, in this one case.

The system checked delivery. It checked formatting. It checked that the merge fields resolved. It checked everything that had been written down as a requirement.

The only check that would have caught the failure was: does this email carry what the agent knows about this family?

Nobody could have run that check from the platform. You could only run it by being a person who had sat in a living room on a Saturday and knew what belonged in the follow up. **The check required the capacity to do the work.**

We had automated the work and left nobody in the loop who could do it.

FIVE QUESTIONS

1. For a task you have automated, what does the system check, and what does it merely deliver?
2. Is there anything the people doing that task were doing that never made it into the process document?
3. How would you find out?
4. If the automation produced a technically correct and substantively empty output, what in your operation would notice?
5. How long would that take?

2. WHAT THEY TOLD US WAS REPEATABLE

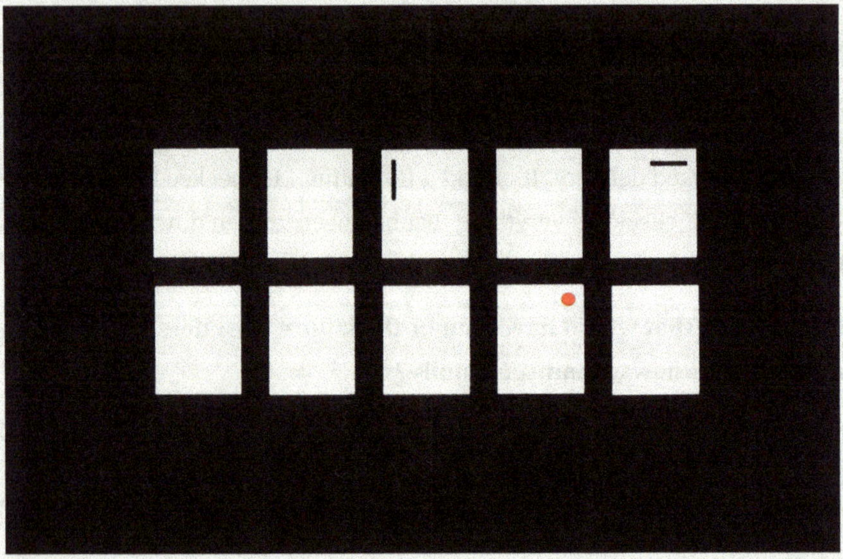

WHY THIS CHAPTER MATTERS

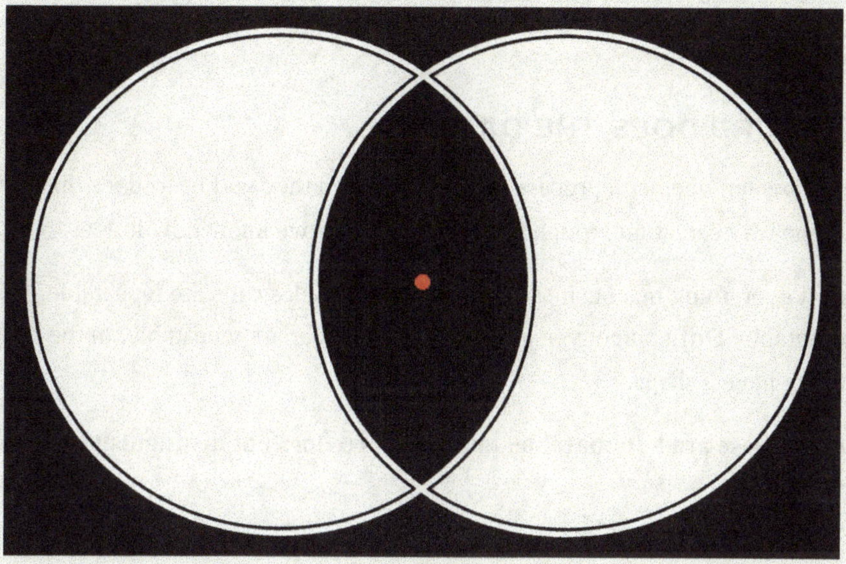

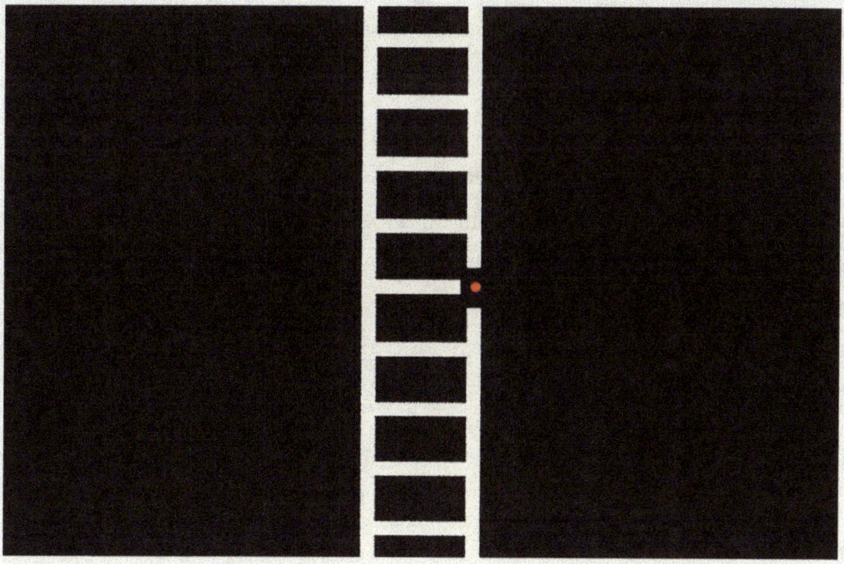

Because "is this repeatable?" is the first question everyone asks and it is almost useless as asked. **The person answering has a different definition of repeatable than you do, and neither of you will discover that until the automation is live.**

THE WORD DOES THE DAMAGE

Ask an owner whether a process is repeatable and they will hear: does this happen the same way each time, roughly, in the sense that we know how it goes.

Almost everything in a business passes that test. Sales calls are repeatable. Hiring is repeatable. Difficult conversations with a supplier are repeatable, in the sense that they have a shape.

None of those are automatable, and the word does not distinguish between them.

WHAT REPEATABLE HAS TO MEAN

For this purpose, repeatable means something much harder:

> ***Given the same input, the correct output is the same, and you can say in advance what it is.***

That is a brutal test and most business processes fail it. The brokerage's follow up sequence failed it. Given the same input, a tour record, the correct output varied enormously, because the correct output depended on ninety minutes of conversation that was not in the record.

HOW TO ACTUALLY ASK

The question is not "is this repeatable." The question is a set of four, and you ask them of the person who does the work rather than the person who owns the business.

One. Show me the last ten times you did this. Not describe. Show. Open the folder.

Two. What is different between them? They will say "nothing." Then you read the ten and you find the differences yourself, and you ask about each one.

Three. When you did the different thing, why? This is where the judgment lives. If they can articulate a rule, the judgment might be encodable. If they say "it just felt like the right thing to say," you have found the edge of the automatable.

Four. What would have happened if you had done it the standard way instead? If the answer is "nothing much," automate it. If the answer is a story, you have found something load bearing.

WHY THE OWNER IS THE WRONG PERSON TO ASK

The founder of the brokerage was not being careless. He was answering from the process as designed, which he knew completely, because he designed it.

The variation lived with the agents, and the agents did not experience it as variation. They experienced it as doing their job.

This is a general pattern and I want to name it clearly. **The further you are from the work, the more repeatable everything looks.** That is not a fact about the work. It is a fact about distance. And the person commissioning the automation is always the furthest away.

THE UNCOMFORTABLE CONSEQUENCE

If the person who does the work has to be interrogated in detail before you can automate it, then the cost of automating is higher than the quotes suggest, and the timeline is longer.

I am not going to soften that. **Discovery done properly on a small automation can take longer than building it.** People hate this and it is why so much of this work goes wrong. The build is the cheap part. The build was never the problem.

WHAT THE TEN EXAMPLES ACTUALLY SHOW YOU

I want to make the audit concrete, because "look at ten examples" sounds like advice and it is a procedure.

You open the last ten outputs. Not a sample chosen by the manager, the last ten in order, including the bad week. You read them side by side rather than in sequence, because differences are invisible when you read one after another and obvious when they sit next to each other.

Then you mark every place where two of them differ, and you do not yet care why. Length, structure, what got included, what got left out, tone, what was attached, who was copied.

Most of what you mark will be noise. **Some of it will be the business.** The only way to tell is to take each one back to the person who produced it and ask what was different about that day, that customer, that order. You will get "I do not know" a few times, and those are fine. What you are hunting is the answer that starts "oh, that one was because."

Every "because" is a rule. Every rule you cannot state is a reason not to automate yet.

I have never run this audit without finding at least one thing nobody had mentioned. At the brokerage it was everything that mattered.

FIVE QUESTIONS

1. When you last asked whether a process was repeatable, who did you ask?
2. Did you look at ten real examples of the output, or did you take a description?
3. What did the person who does the work say was "nothing different" that turned out to be different?
4. Can you state, in advance, what the correct output is for a given input?
5. If not, what would you need to learn to be able to?

3. THE THREE VARIABLES

WHY THIS CHAPTER MATTERS

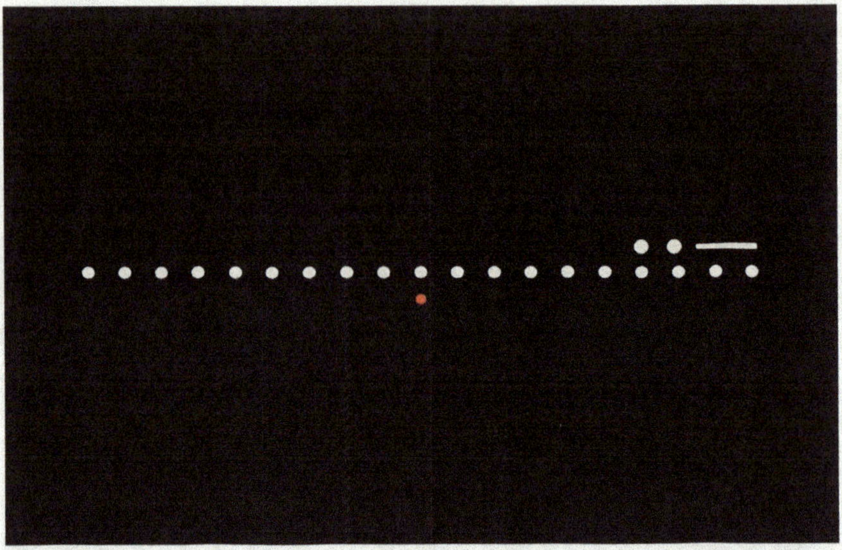

Because this is the gate, and it is the shortest useful thing in the book. Three conditions, all of which must hold. **Most of what businesses try to automate fails at least one and gets automated anyway.**

THE THREE

A task is a candidate only if all three are true.

One. It is repeatable, in the hard sense from chapter two. Same input, same correct output, statable in advance.

Two. It requires almost no customisation. Not "the customisation can be handled by branching." Almost none.

Three. It takes almost no thought. If the person doing it has to consider anything, you are looking at judgment, and judgment is the thing that does not survive the transfer.

And a fourth statement, which is the test of the three rather than a condition of its own:

> ***It can be handed to someone else and it will not break.***

If you cannot hand the task to a competent stranger with the written procedure and get your output, you do not have a task. You have a person.

THE MISTAKE EVERYONE MAKES

Here is the thing I see more than any other, and it is close to universal.

> ***People do not automate the tasks that qualify. They automate the tasks they hate.***

The request almost always arrives in this form: I want to automate this so I do not have to do it any more.

Notice what that sentence is about. It is about the person's relationship to the work, not about the work's properties. And the tasks people hate are almost never the simple repetitive ones. **They hate the complicated ones. The ones with judgment in them, and exceptions, and awkward customers, and consequences when they get it wrong.**

Those are exactly the tasks that fail all three variables.

So the pattern is not that businesses automate badly. It is that **the desire to automate is inversely correlated with eligibility to automate**, and nothing in the sales process for any tool anywhere corrects for it.

WHAT PASSING THE GATE ACTUALLY FEELS LIKE

An eligible task is boring. It is so boring that you will be slightly embarrassed to be paying for the automation.

That is the correct feeling. The satisfying automation, the one that makes you feel you have finally got a handle on something difficult, is the one that will be wrong in March and nobody will know.

AND A NOTE ON SCALE

The three variables do not scale with your business size. I have applied them to a business making a thousand dollars a month and to companies in the Fortune 500. The tasks are different sizes. **The gate is identical**, because the gate is about the structure of the work and not the volume of it.

WHAT TO DO WHEN A TASK FAILS ONE VARIABLE

Failing the gate is not the end of the conversation, and treating it as a yes or no wastes most of the value in the assessment.

If it fails repeatability, the question is whether it fails because the process is genuinely variable or because it was never written down. Those look identical from the outside and they have opposite remedies. One is a discovery problem you can fix in a fortnight. The other is a property of the work.

If it fails on customisation, look at where the customisation enters. Frequently it enters at one step out of nine, which means eight steps are eligible and one is not, and you have just found the correct shape of the automation rather than a reason to abandon it.

If it fails on thought, stop. That is the one that does not decompose. A task requiring judgment is not a task with a judgment step you can carve out, because the judgment is usually distributed across the whole thing and the person doing it cannot tell you where it lives.

Two failures out of three and I do not take the engagement. Not because it cannot be built. Because it will be unchecked within a year and I will have sold somebody a liability with an invoice attached.

FIVE QUESTIONS

1. Take the last thing you automated. Which of the three variables did it actually pass?
2. Was the request "this is eligible" or "I do not want to do this any more"?
3. What is the most boring task in your operation, and is it automated?
4. What is the task you most want to stop doing, and does it pass the gate?
5. If those are different tasks, which one are you funding?

4. A TASK, NOT A JOB

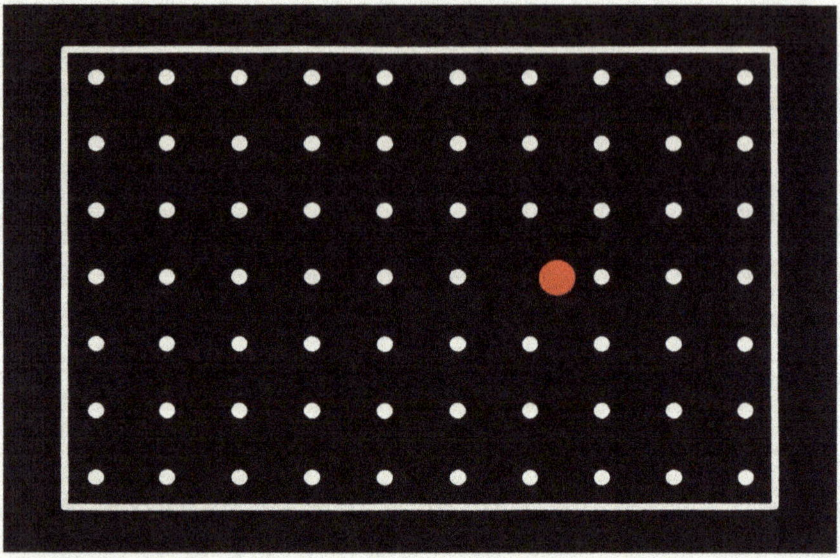

WHY THIS CHAPTER MATTERS

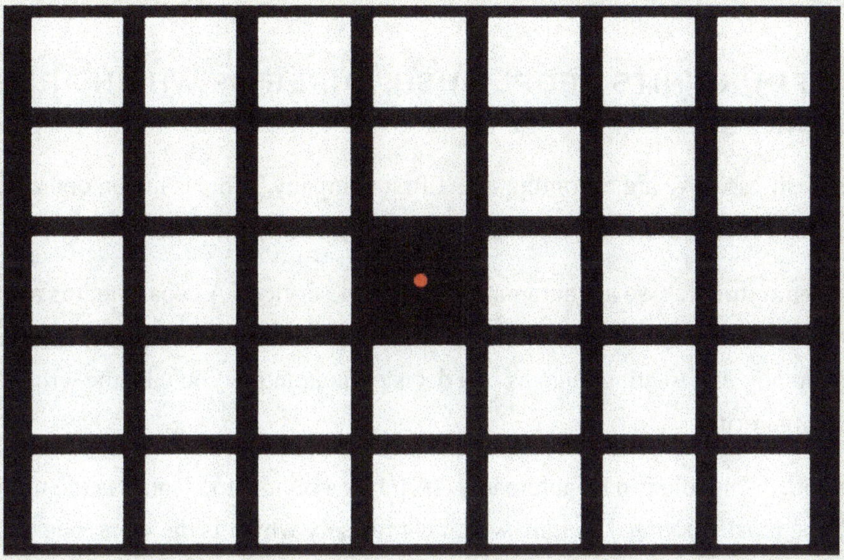

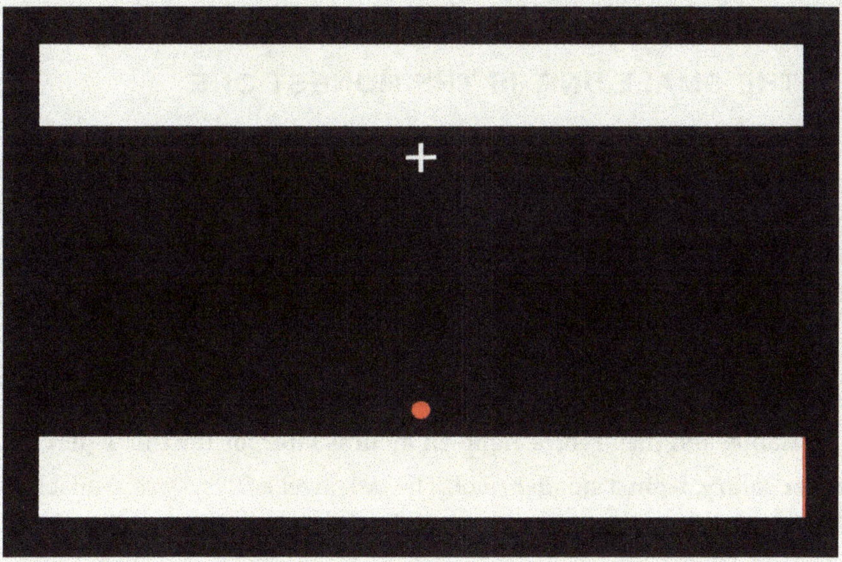

Because the unit is wrong in almost every conversation about this, and once the unit is wrong every downstream decision is wrong. **You cannot automate a job. You can only ever automate a task.**

THE FOUR UNITS PEOPLE USE, THREE OF WHICH ARE MISTAKES

The business. "We are becoming an AI first company." This is not an operational statement. It cannot be assessed, checked, or reversed. Ignore it.

The department. "We are automating customer service." A department is a bundle of dozens of tasks with wildly different properties. Some pass the gate, most do not, and treating them as one decision guarantees you take the wrong one for the majority.

The job. "This role can be automated." A job is a bundle too, and it is the unit with the most emotional weight, which is precisely why it is the worst one to reason with. Chapter fifteen deals with what happens when you get this wrong.

The task. The smallest complete unit of work that produces an output somebody else uses. **This is the only unit that can be assessed.**

WHY THE SMALL UNIT IS THE HONEST ONE

When you evaluate at task level, something useful happens. You discover that a role people describe as "automatable" is usually about sixty percent eligible tasks and forty percent judgment, and that the forty percent is why the role exists.

That finding is available only at the small unit. At job level it is invisible, and the decision collapses into a yes or a no about a person.

I have watched businesses eliminate roles that were seventy percent automatable, and then rehire eighteen months later for the thirty percent, at a higher salary, from a smaller pool. The task level analysis was available before the decision. Nobody ran it, because the question had been asked at the wrong altitude.

If you read Volume IV, this will sound familiar. It is the same defect in a different object. **The altitude a decision is made at determines what is visible when you make it.**

THE PRACTICAL CONSEQUENCE

Your automation roadmap should be a list of tasks, and it should look unimpressive.

Not "automate onboarding." **"Automate the welcome email, the account provisioning request, and the calendar invite for the kickoff call."** Three items, three separate gates, three separate checkers, three separate reversal tests.

The remaining eleven things in onboarding stay manual, and you can now say precisely why each one does, which is worth more than the automation.

THE EXCEPTION, AND IT IS NOT REALLY ONE

Occasionally somebody points out that a chain of eligible tasks can be automated end to end, and that this looks like automating a process.

It does look like that, and the distinction still matters, **because each link keeps its own check and its own checker.** A chain of six automated tasks with one checker at the end is not six automations. It is one automation with five unchecked steps inside it, and it will fail at whichever step nobody is watching.

HOW TO BUILD THE INVENTORY

The task list is the deliverable nobody wants to pay for and the one that determines everything downstream.

Sit with the person doing the work for a full cycle. A day, a week, whatever the natural period is. Write down every distinct unit of work that produces an output somebody else uses. You are looking for the seams, the places where they finish one thing and pick up another.

You will get somewhere between fifteen and sixty items for a single role. **That number surprises people every time**, and the surprise is the point: the role was being discussed as one decision and it is forty.

Then three columns. **Passes the gate. Fails the gate. Do not know yet.**

The third column is the important one and most people will not allow themselves to have it. It is where the honest uncertainty goes, and a task in that column gets the ten example audit before it gets a budget.

A business that has this inventory can have a sensible conversation about automation. A business without one is guessing at altitude, which is Volume IV's finding in different clothing.

FIVE QUESTIONS

1. When you last discussed automation, what unit was the conversation held at?
2. Take a role you have described as automatable. List its tasks. How many pass the gate?
3. What percentage of that role is the part that does not pass?
4. Does your roadmap contain tasks or does it contain departments?
5. For your longest automated chain, how many separate checks exist inside it?

PART II

THE CHECK

5. YOU CANNOT CHECK WHAT YOU COULD NOT HAVE DONE

WHY THIS CHAPTER MATTERS

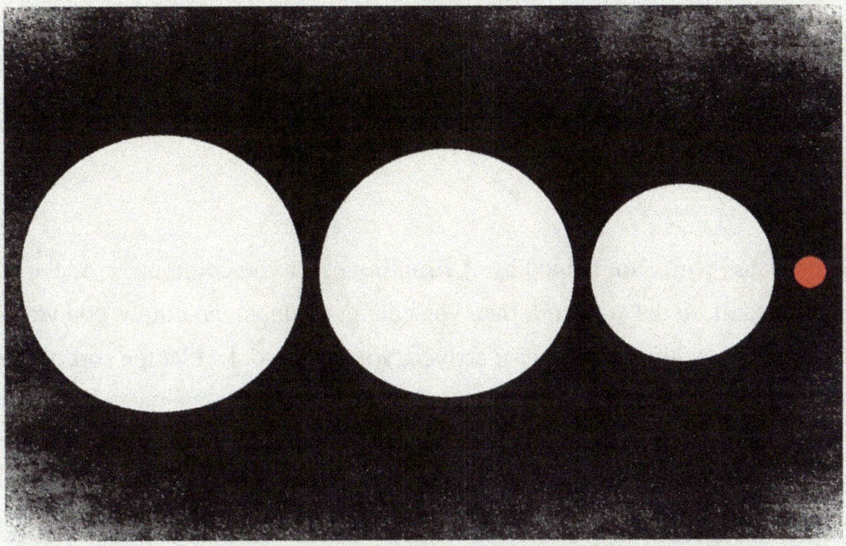

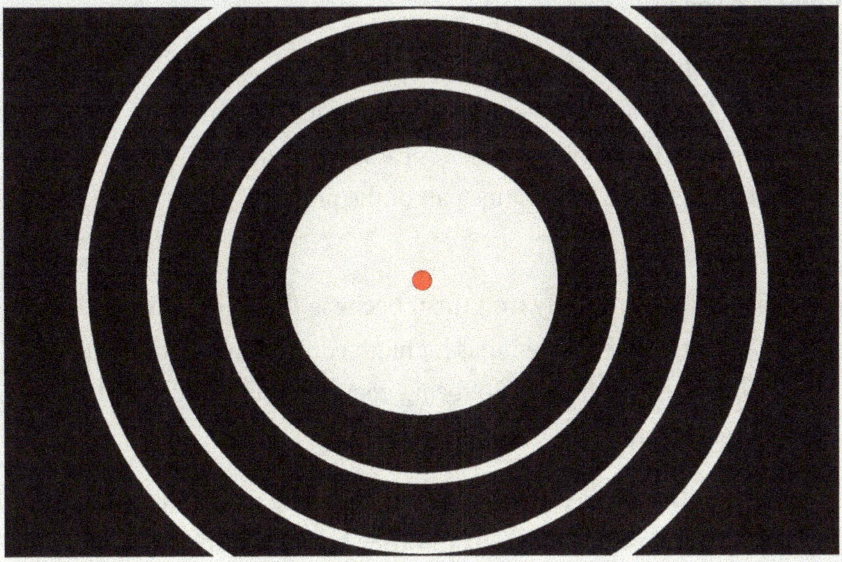

Because this is the claim the rest of the book rests on, and once you have seen it you will not be able to stop seeing it in your own operation.

THE CLAIM

> ***The ability to supervise a piece of work is a function of the ability to do that work.***

Not correlated with. Not helped by. **A function of.** If you could not produce the output yourself, to the standard, then you cannot evaluate the output you were given. You can evaluate whether it arrived, whether it looks like the sort of thing you expected, and whether it is formatted well. You cannot evaluate whether it is right.

HOW I KNOW

I write software and I design systems. When I use a current large language model to write code, I read what it produces, and **more often than not I find something in it that is wrong.**

I want to be careful here. That is my experience of my own work, not a measurement. I have not counted, and if I gave you a percentage I would be inventing it, which is the thing this series exists to argue against. What I can tell you is that finding errors is a routine part of the process rather than an occasional event.

Now the important part. **I only find them because I can write the code myself.** The errors are not flagged. Nothing is highlighted. The output looks finished, because it is finished, in the sense of being complete and well formed. What it is not, sometimes, is correct.

A person who cannot write the code receives the same output. It looks identical to them. **It looks finished to them too, and they have no mechanism at all for discovering otherwise, so they ship it.**

THE LINE I USE WITH CLIENTS

> ***A current large language model can teach you how to perform heart surgery. That does not mean you should perform heart surgery.***

People laugh, and then about a third of them stop laughing, because they recognise something.

The distinction people miss is between **access to instruction** and **capacity to evaluate**. Those have never been the same thing, and the gap between them has never been this easy to ignore. You can now obtain a competent looking answer in any domain on earth in four seconds. Your ability to tell whether it is a good answer has not improved at all.

WHERE THIS ACTUALLY BITES IN A BUSINESS

Not in the dramatic places. In the ordinary ones.

A founder with no finance background asks for a cash flow model and gets one. It has the right rows and it looks like a cash flow model. Whether the working capital assumptions are sane is a question they cannot answer, and the model will now inform decisions for two years.

An operator with no legal training generates a contract clause. It reads like a contract clause. Whether it does what they need in the jurisdiction they are in is unknowable to them.

A marketing lead with no analytics background gets an attribution analysis. It has numbers, and charts, and a conclusion.

In every case the output is plausible, complete, and confidently presented, and the person receiving it has no independent standard to hold it against. They will treat it as done, because from where they stand it is indistinguishable from done.

THE COUNTERARGUMENT, AND MY ANSWER

Somebody always says: by this logic nobody could ever delegate anything.

That is not right, and the difference is worth being precise about. When you delegate to a person, you are not only relying on your ability to check. You are relying on the person's stake, their reputation, their fear of being wrong, their willingness to say "I am not sure about this bit." **Those are all checks that operate independently of your competence.**

A system has none of them. It does not hedge unless asked. It does not flag its own uncertainty in a way that maps to actual risk. It will not come back on Thursday and say it has been thinking about it and something is bothering it.

Delegating to a person imports their judgment. Delegating to a system exports yours. If you have none to export in that domain, the arrangement has no judgment in it anywhere, and nothing about the output will tell you that.

FIVE QUESTIONS

1. What have you automated in a domain where you could not do the work yourself?
2. Who in your business could do it?
3. Have they looked at the output?
4. What would a subtly wrong output in that domain look like, and would you recognise it?
5. If the answer to four is no, what is currently standing between you and a bad decision?

6. THE GOLD STANDARD PROBLEM

WHY THIS CHAPTER MATTERS

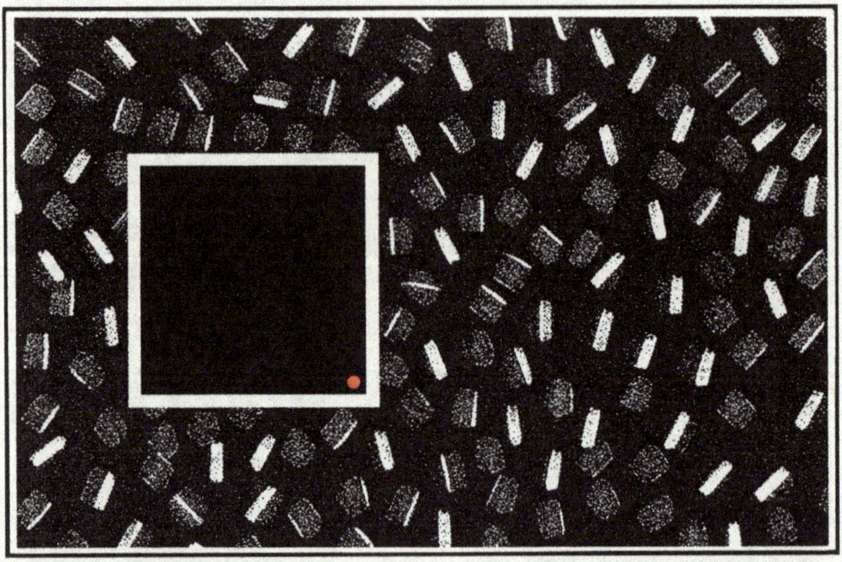

Because the failure mode is not that people are fooled. It is that they stop asking. **Output from a machine gets a level of deference that no human colleague has ever received**, and that deference is the mechanism by which unchecked work enters the business.

WHAT I SEE

People treat what a system tells them as true.

Not provisionally true, not a good first draft, not a starting point. **True.** They take the output as the gold standard and they do not challenge it, and when I ask why, the honest answer is usually that they would not know how to challenge it.

This is worse than ordinary overconfidence, because it is self reinforcing. The less able you are to evaluate the output, the more impressive it looks, and the more impressive it looks the less inclined you are to try.

THE TWO MOVES THAT CREATE IT

Move one. You ask for something you could have done yourself. You read it, you notice it is decent, you fix two things. Healthy. This is the good case and it is where most people start.

Move two. You ask for something you could not have done yourself. It arrives looking exactly like the output from move one. Same fluency, same confidence, same structure.

You have no way to tell move one from move two by looking at the result. The interface does not change. The tone does not change. The only thing that changed is the thing you cannot see, which is whether you are equipped to evaluate what you just received.

And because move one went well, you extend the trust you earned there into a domain where you did not earn it.

IT IS NOT A TECHNOLOGY PROBLEM

I want to be careful not to blame the tool for this, because it is not the tool's doing.

The systems are not claiming certainty. Most of them hedge constantly, and users find the hedging irritating and ask them to stop. **The deference is coming from the human side of the arrangement.** We built a thing that produces fluent, complete, well organised output on demand, and fluency has always been the signal humans use for competence, because until recently fluency was expensive.

It is not expensive any more. **The signal has come apart from the thing it used to indicate**, and our instincts have not been updated, and they are not going to be updated by anyone telling you about it, including me.

WHAT ACTUALLY CORRECTS IT

Not scepticism. Scepticism is a mood and it does not survive a busy week.

The only thing that corrects it is a standard that exists outside the output.

If you have a written description of what a correct result looks like, produced before you saw this result, then you have something to compare against and your feelings about the output are irrelevant. If you do not, then you are evaluating the output against your impression of the output, which is not a check. It is a mirror.

This is why chapter three insists the task have a written procedure before it is automated. The procedure is not bureaucracy. It is the only artefact in the whole arrangement that is not downstream of the system you are trying to check.

THE DEMONSTRATION PROBLEM

There is a specific place this failure gets manufactured, and it is worth naming because it is where most businesses form their first impression.

A demonstration is built to be checkable by someone who cannot check.

It uses a clean example. The input is well formed, the domain is one everybody in the room understands, and the output is impressive precisely because the audience can evaluate it. That is good salesmanship and it is not dishonest.

The problem is what the audience concludes. They watched themselves successfully evaluate an output, and they generalise from that to a belief that they will be able to evaluate outputs. What they actually demonstrated was that the tool performs well on a task they happened to be competent at.

Then it goes into production against messy inputs in a domain where nobody in the building has the underlying expertise, and the confidence formed in the demonstration comes along unchanged.

When you evaluate a tool, use your worst real input in your least understood domain, and have the one person who knows that domain grade it. If that changes your opinion, the demonstration was measuring the audience.

FIVE QUESTIONS

1. When did you last reject an output from an automated system and require it to be redone?
2. On what grounds?
3. Was that ground written down before you saw the output?
4. In which domains do you accept output without challenge, and is that because it is reliably good or because you could not mount a challenge?
5. Who in your business is willing to say "this is wrong" about an automated output, and what happens to them when they do?

7. WHAT THE OPERATOR LOSES BY NOT DOING THE WORK

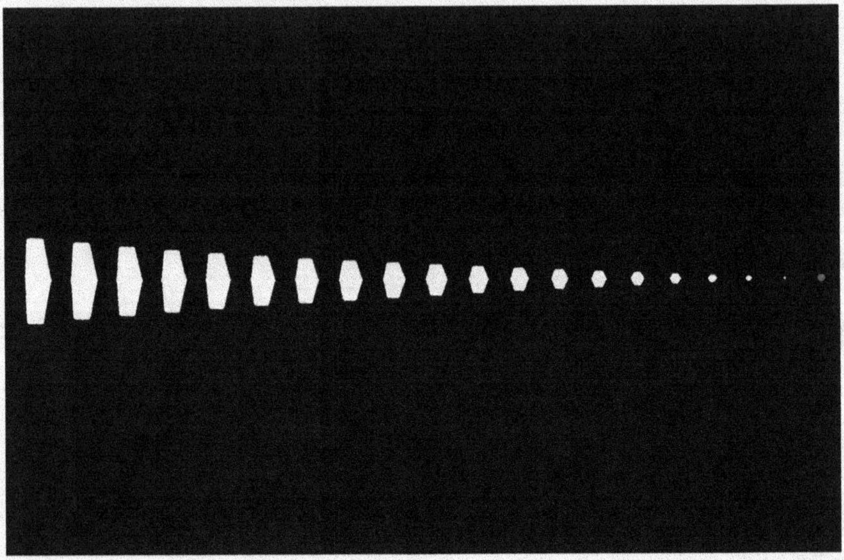

WHY THIS CHAPTER MATTERS

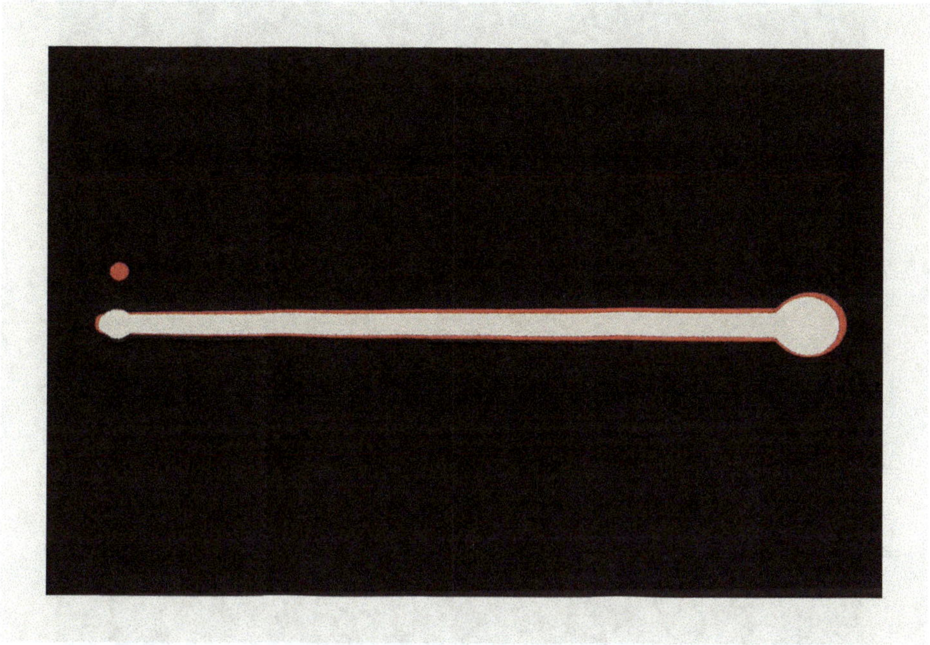

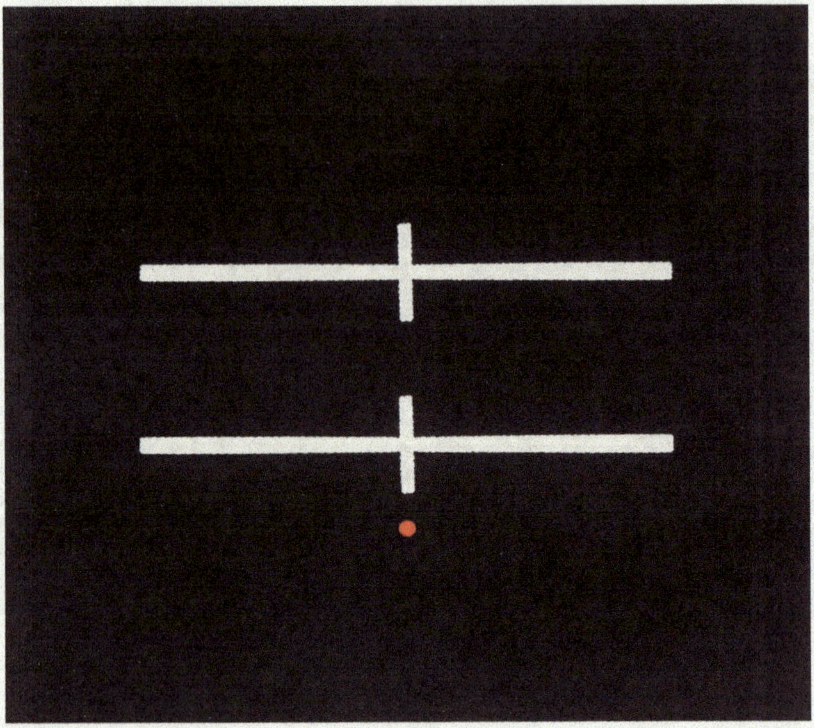

Because everything up to this point has been static, and it is not static. **The capacity to check does not just have to exist. It has to be maintained, and automating the task is what stops it being maintained.**

WHERE THIS CAME FROM, AND IT WAS NOT FROM ME

I need to be straight about the provenance of this chapter, because the rest of the book is my own experience and this part is not.

When this volume was being researched, my position was the one in chapter five: the person who can do the work can check it, the person who cannot, cannot. Static. A property of people.

Then the research turned up a paper I had never read, from a field I have never worked in, published in 1983. It is called *Ironies of Automation*, it was written by Lisanne Bainbridge, and it appeared in a control engineering journal called *Automatica*. It is about process plants. Chemical works, power stations, the kind of place with a control room.

It says what I said, and then it says the thing I had missed, and the thing I had missed is more important than the thing I had right.

I am telling you this because a book that presents its author as having known everything in advance is not a book about operations. Twenty years of practice got me to a true observation. It did not get me to the mechanism, and somebody got there four decades earlier in an industry I have never set foot in.

THE FIRST IRONY

> ***"The designer who tries to eliminate the operator still leaves the operator to do the tasks which the designer cannot think how to automate."***

Read that again with your own operation in mind.

Whatever you automated, you automated the part you could specify. What is left with the human is, by definition, **the residue: the part nobody could work out how to encode.** That residue is not the easy remainder. It is the hardest fragment of the original job, stripped of all the routine work that used to surround it and make it comprehensible.

We describe this as freeing people up for higher value work. Sometimes that is exactly what it is. But notice that it happened by subtraction rather than design, and that nobody checked whether the residue was a coherent job.

THE SECOND IRONY, WHICH IS THE ONE THAT MATTERS

> ***"Physical skills deteriorate when they are not used, particularly the***

> ***refinements of gain and timing."***

This is why chapter five is not enough.

The person who could check the work loses the ability to check the work, precisely because the automation is running. Not through neglect. Not because they got lazy. Because skill is maintained by exercise and the automation removed the exercise.

So the arrangement contains a decay term nobody put in the business case. On day one you have an eligible task, a written standard, and a named person who can reverse engineer the output. Three years later you have the same task, the same standard, the same person, and **that person can no longer do the work to the standard, because they have not done it for three years.**

Nothing changed. Nothing broke. Nothing appears in any report. **The check quietly stopped being a check and became a ritual.**

THE THIRD IRONY, AND IT IS THE CRUEL ONE

> ***"When manual take-over is needed there is likely to be something wrong with the process, so that unusual actions will be needed to control it, and one can argue that the operator needs to be more rather than less skilled, and less rather than more loaded, than average."***

The moment a human has to step in is, by definition, the moment the situation is abnormal. Automations do not need rescuing when things are going well.

So the person taking over is facing the hardest version of the task, **using skills they have not practised since the automation went in, under time pressure, with the business already degrading.** And we staff that position with whoever is left after the efficiency saving.

Bainbridge's conclusion is that the supervising operator needs to be **more skilled than average and less loaded than average.** Not less. More.

I have never once seen a business staff an automated process that way. I have never proposed it. Reading that sentence changed how I scope engagements.

AND THE FOURTH, WHICH EXPLAINS THE BROKERAGE

> ***"It is impossible for even a highly motivated human being to maintain effective visual attention towards a source of information on which very little happens, for more than about half an hour."***

Monitoring is the task humans are worst at, and monitoring is exactly what automation leaves us.

This is not a motivation problem and you cannot fix it with a policy. Attention to an uneventful signal collapses in about half an hour, in motivated people, under laboratory conditions. Your team is not going to beat that by caring more.

Which is why the check has to be an event, not a state. Nobody watches an automation. Somebody, on a schedule, takes an output and works it backwards. That is a task with a beginning and an end, and humans can do those.

WHAT THIS MEANS FOR THE GATE

Chapter three gave you three variables. They are necessary and they are not sufficient, and here is the fourth condition, which is Bainbridge's rather than mine:

> ***The person who checks must still be doing enough of the work to remain able to.***

Which means an automation is not a decision you make once. **It is a position you hold**, and holding it costs something every year, and if you stop paying, the check expires without notifying you.

FIVE QUESTIONS

1. For your oldest automation, who checks it, and when did they last do the task by hand?
2. Is there anything in your operation that would tell you their capacity had decayed?
3. What is the residue left with humans after your largest automation, and is it a coherent job?
4. If your most important automation failed tomorrow, who takes over, and are they your most skilled person or your most available one?
5. Is your check an event on a schedule, or is it somebody being expected to notice?

8. CONFIDENCE IS THE FIRST THING TO FAIL

WHY THIS CHAPTER MATTERS

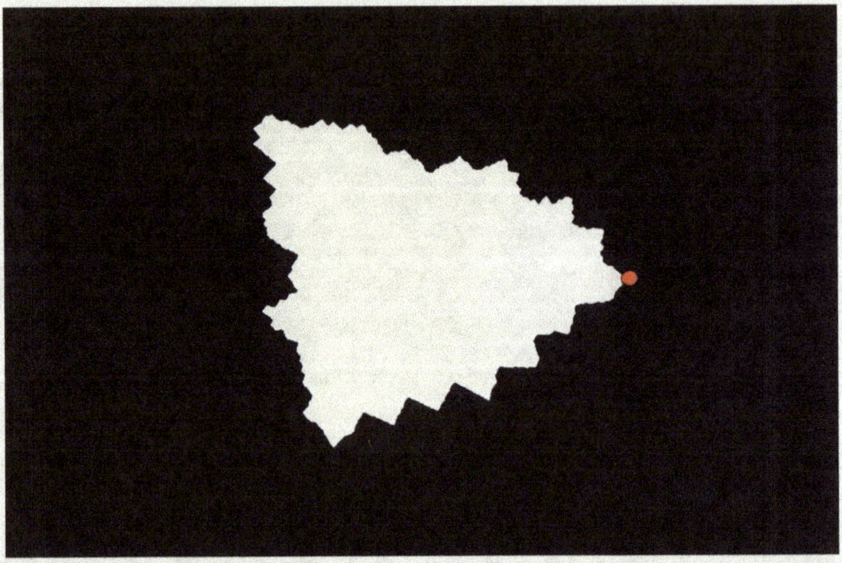

Because you might have read chapter five and concluded that expertise protects you. It does not. **It protects you from one failure and leaves you fully exposed to a second one, and the second one has better evidence behind it.**

TWO STUDIES

I am going to give you two pieces of research, and then the thing they mean together, which neither says on its own.

THE FIRST. PEOPLE WROTE WORSE CODE AND FELT BETTER ABOUT IT.

A controlled study presented at a major security conference in 2023 gave participants security related programming tasks. Some had an AI assistant, some did not.

The finding was in two halves.

> ***Participants with the assistant wrote significantly less secure code than those without.***
>
> ***Participants with the assistant were more likely to believe they had written secure code.***

Same people. Same study. **Output got worse and confidence went up.**

I should say plainly that the model in that study is now several generations old, and I am not offering this as a claim about any current system's capability. **It is a finding about the human relationship to the tool**, and that is the part that has not changed, because the thing that changed is the model and the thing under examination is us.

THE SECOND. EXPERTS GOT SLOWER AND COULD NOT TELL.

In 2025 a research group ran a randomised trial with sixteen experienced developers working on large open source repositories they already knew well. Codebases with over a million lines. People with hundreds of hours of experience using these tools.

These are not naive users. **These are precisely the people chapter five says should be able to check.**

They were given real tasks, some with AI tools allowed and some without.

What they predicted before starting	**24 percent faster**
What was measured	**19 percent slower**
What they believed afterwards, having just done it	**20 percent faster**

Read the third row again. **They did the work. They experienced the slowdown. And then they reported that it had sped them up by twenty percent.**

The researchers are careful about scope, and so am I: they state explicitly that they do not claim their developers or repositories represent most software work. This is one study, on expert developers, on familiar code. It is not a general law.

It is also the single most uncomfortable result I encountered while writing this book, because these were my people, doing my kind of work, and they were wrong about their own experience in the direction of optimism.

WHAT THE TWO MEAN TOGETHER

They are measuring different things and that is the point.

The first study is about correctness. Whether you can tell right from wrong. Expertise governs that, which is chapter five, and chapter five stands.

The second study is about the arrangement. Whether the whole setup is actually working. And expertise does not govern that at all. **The experts were as**

wrong as anyone, in the same direction, after direct personal experience of the opposite.

So:

> *The check cannot be a feeling. It has to be an instrument.*

Your sense that the automation is working is not evidence that the automation is working, and your seniority does not upgrade that sense into evidence. **The people best equipped to check are also the people most confident that everything is fine, and those two facts turn out to be the same fact**, because confidence is what expertise feels like from the inside and it does not switch off when it stops being warranted.

WHAT TO DO ABOUT IT

Measure the outcome, not the experience. Not "is this working" but "what was the number before and what is it now." The brokerage found their failure because somebody was counting opens. They did not find it by asking whether the sequence felt effective. Had they asked, everyone would have said yes, because the emails were going out.

Have somebody who was not involved run the check. Not because your team is dishonest. Because the person who built it has the most confidence and the least ability to see the arrangement.

Write down what you expect before you start. The developers in that study forecast 24 percent faster. Because that forecast existed, the gap could be measured. **A prediction you did not write down cannot be wrong**, which is comfortable, and useless.

FIVE QUESTIONS

1. For your last automation, what did you predict would improve, in numbers, before it went in?

2. Where is that written down?
3. What actually happened to that number?
4. If nothing was measured, on what basis do you currently believe it worked?
5. Who checks your automations, and were they involved in building them?

9. DRIFT, AND WHY NEITHER SIDE CAN SEE IT

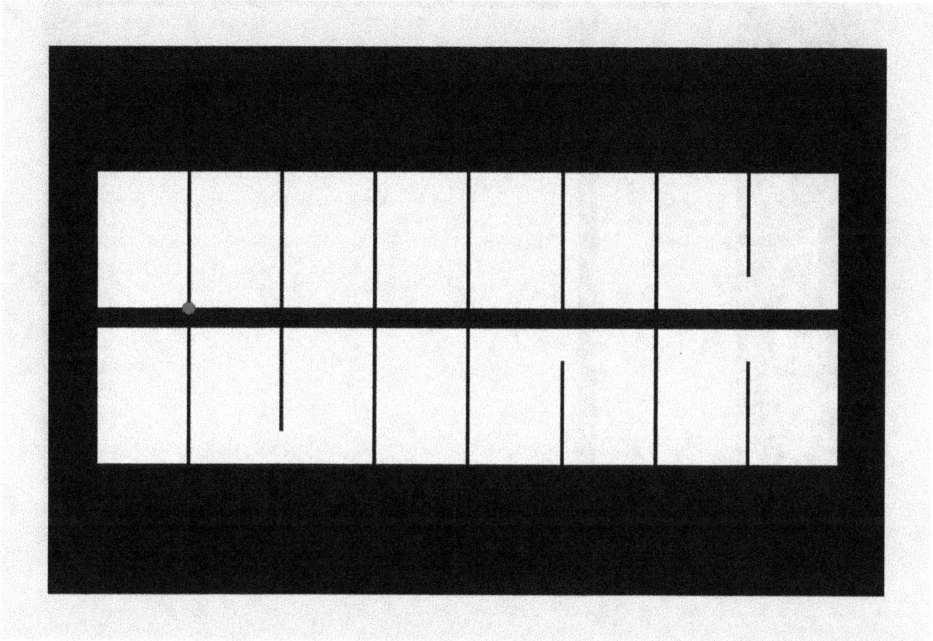

WHY THIS CHAPTER MATTERS

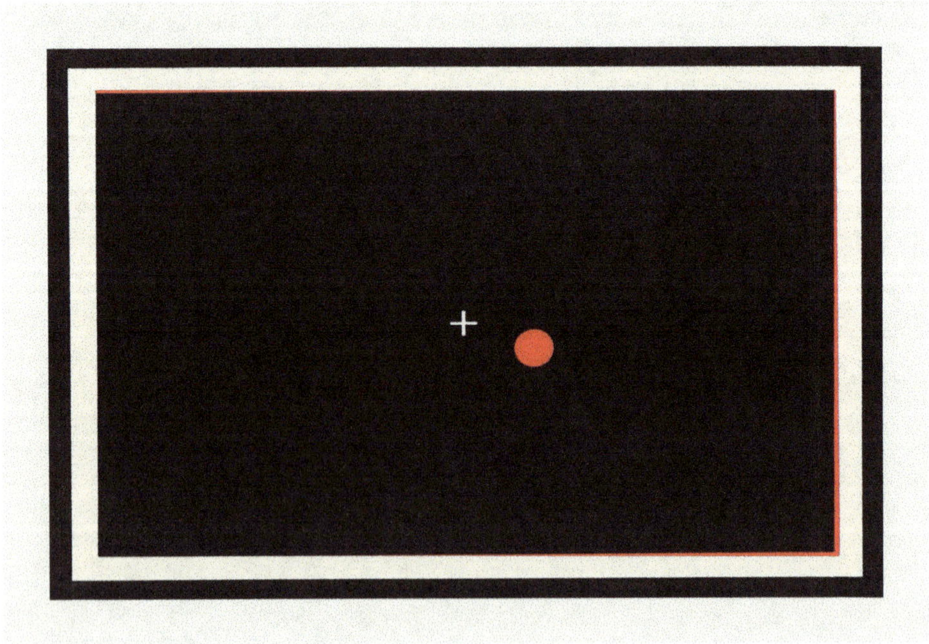

Because everything so far has been about the moment of automation. **Drift is what happens afterwards, over months, and it is invisible from both ends of the arrangement at the same time.**

WHAT DRIFT IS

The task is still running. The output is still being produced. And slowly, the output stops being right.

Sometimes the world moved. Prices changed, the product changed, the regulation changed, the customer changed. Sometimes the inputs degraded. Sometimes the system was updated by somebody else and behaves slightly differently now.

Nothing announces any of this. The automation continues to do what it does, competently, at volume.

WHY THE HUMAN CANNOT SEE IT

Because of everything in chapters five through eight. If they could not do the work, they never could check. If they could, they have not practised since it went in. And their confidence in the arrangement is not tracking its actual performance.

Add the ordinary fact that the automation was purchased to stop somebody thinking about this, and had it stayed on anyone's mind it would have failed its own business case.

WHY THE SYSTEM CANNOT SEE IT EITHER

Here is the part people find genuinely surprising.

> ***The system does not know what drift is, because nobody ever defined correct for it.***

We talk about systems detecting anomalies. What they detect is deviation from a pattern in what they have seen. **That is not the same as deviation from correct.**

If the output has been subtly wrong since March, the wrongness is the pattern, and by April it is the baseline.

The brokerage's platform could report delivery, opens and clicks. It could not report "these emails no longer contain what makes them work," because that was never expressible in the system. **It had no representation of the thing that mattered.** Not through poor design. Because nobody had ever written it down anywhere, including the people doing it.

THE SYMMETRY

> ***The human cannot see the drift because they no longer hold the standard. The system cannot see the drift because it was never given the standard.***

Neither party in the arrangement holds it. Both are behaving reasonably. Both would tell you everything is fine. And they would agree with each other, which is the most dangerous property of the whole configuration, because agreement between two parties who are both unable to see is indistinguishable from confirmation.

WHAT ACTUALLY CATCHES IT

A standard written outside the system, and a person who periodically works backwards from the output to that standard.

Backwards is not decoration. Anyone can look at an output and ask whether it looks right, and that check is worthless, because it compares the output to your impression of the output. **Working backwards means taking what was produced and reconstructing what the input must have been and what should have happened, then comparing that against what did.**

The gaps show up in the reconstruction. That is the only place they show up.

HOW OFTEN

More often than you want, less often than continuously.

The honest answer is that it depends on how fast your inputs move, and most businesses have never asked. **Pick an interval, put it in a calendar, give it to a named person, and treat a missed check as an incident rather than a scheduling problem.** If the check is not on someone's calendar it is not happening, whatever your process document says.

WHAT DRIFT LOOKS LIKE IN PRACTICE

An example, because the abstract version is easy to nod at and hard to recognise.

A business automates the routing of inbound enquiries. Rules based, sensible, tested. Enquiries mentioning certain terms go to the specialist team, everything else goes to general handling. It works.

Fourteen months later the specialist team is quiet and general handling is drowning, and everybody assumes enquiry mix has shifted.

What actually happened is that the marketing team changed the language on the website. Customers now describe the same problem using different words, because they are copying the words they were given. The router is applying its rules perfectly to a vocabulary that no longer exists.

Nothing broke. The router reports healthy. Every enquiry was routed. **The specialist team's silence is the only symptom, and it reads as good news.**

Notice the shape. The change came from a different department, was correct on its own terms, and nobody connected it to the automation because nobody was holding both things at once. **Drift usually arrives from outside the system that suffers it**, which is why monitoring the system will not catch it and reconstructing the output will.

FIVE QUESTIONS

1. For each of your automations, when was the output last reconstructed backwards rather than glanced at?
2. Is there a written standard for correct that predates the automation?
3. If the output had been subtly wrong for four months, what would have surfaced it?
4. Who owns the check, and is it in their calendar?
5. What has changed in your business since the automation went in that the automation does not know about?

PART III

THE TOOL

10. THE TOOL AND THE TASK

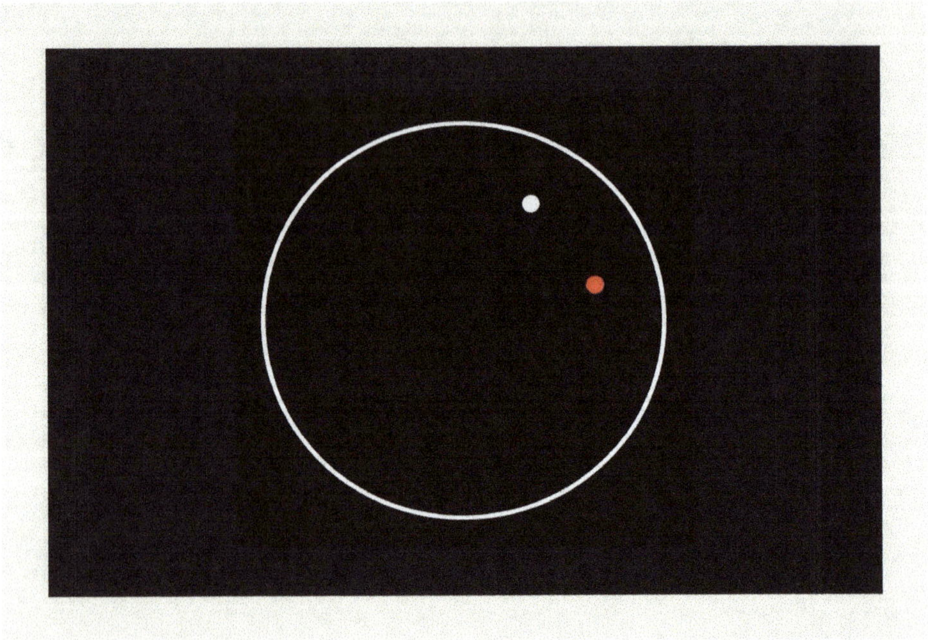

WHY THIS CHAPTER MATTERS

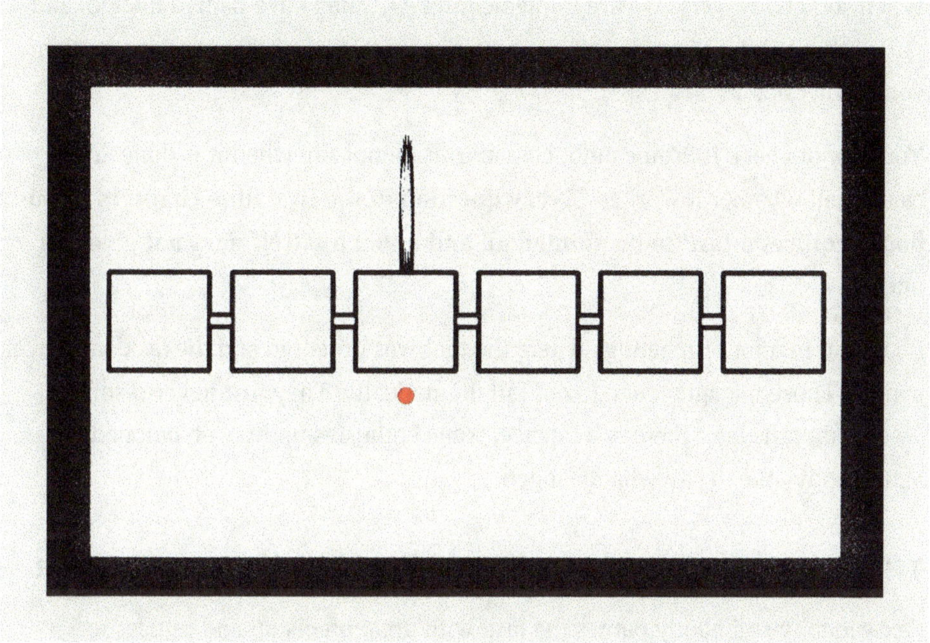

Because two words are used interchangeably in every conversation about this and they are not the same thing. **Getting them confused is why businesses buy a capability and expect an outcome.**

THE DISTINCTION

> ***AI is the tool. The automation is the task.***

AI is a product. A universe of possible uses. It can search, draft, summarise, classify, transcribe, translate, extract, generate, converse. That is not a list of things it does for you. It is a list of things it could be pointed at.

The automation is one specific thing you have pointed it at. If AI is the universe, the automation is a single moon: one task, one input, one expected output, one place it sits in your operation.

WHY THE CONFUSION IS EXPENSIVE

When somebody says "we are implementing AI," they have named the tool and said nothing about the task, and the sentence therefore contains no information you can act on.

You cannot check it. You cannot cost it. You cannot say when it is done. You cannot tell whether it worked. **Every question in the first nine chapters of this book requires a task to be pointed at, and a tool by itself does not give you one.**

I have sat in a lot of meetings where the tool was procured and the task was never named. Those engagements do not fail dramatically. They just never resolve. Twelve months later there is a licence, some enthusiasm, three abandoned pilots, and nobody able to say what changed.

THE REVERSE ERROR, WHICH IS RARER AND CHEAPER

Occasionally somebody names the task with great precision and refuses to consider whether the tool fits. That is a smaller mistake because it fails fast. You try, it does not work, you stop.

The tool-without-task error does not fail fast. It fails slowly and expensively, because there is no criterion for stopping.

WHAT "AI CAN DO ANYTHING" ACTUALLY MEANS

It is a claim about the category, and it is roughly true about the category.

It is not a claim about the specific system in front of you, configured the way you have configured it, on your data, for your task. Those are different questions and only the second one matters to your operation.

The category can do almost anything. The instance in your business can do the things you have set it up to do and verified, and no others, and the gap between those two statements is where most of the disappointment in this field lives.

WHAT TO WRITE DOWN BEFORE YOU SPEND ANYTHING

One sentence, in this shape:

> ***"We are using [a tool] to do [one named task], which produces [a named output], checked by [a named person] against [a written standard], and we expect [a named number] to change."***

If you cannot complete that sentence, you are not ready, and no amount of tooling will make you ready, because the missing parts are not technical.

Every clause maps to something earlier in this book. The named task is chapter four. The written standard is chapter six. The named person is chapters five and seven. The named number is chapter eight.

THE PILOT THAT NEVER ENDS

The clearest symptom of a tool procured without a task is the pilot with no end date.

A pilot is supposed to be a test, and a test needs a hypothesis, a duration and a decision rule. **What I usually find is a pilot that has been running for nine months, has no criterion for success, and is described as going well.**

It is going well because nobody has defined what going badly would look like. The team is enthusiastic, there are anecdotes, somebody saved an afternoon. None of that is evidence and all of it feels like evidence.

Ask three questions and it resolves immediately. What did you expect to change, in a number. What is that number now. What would have to be true for you to stop.

If the third question produces a long pause, the pilot is not a test. **It is a subscription with a story attached**, and it will run until somebody audits the budget.

FIVE QUESTIONS

1. Write the sentence above for something you are currently building. Which clause did you struggle with?
2. Has anyone in your business named a tool and called it a strategy?
3. What can you actually verify your system does, as configured, on your data?
4. What are you assuming it does because the category can?
5. Which of your projects has no stopping criterion?

11. THE SHOVEL AND THE FLATHEAD

WHY THIS CHAPTER MATTERS

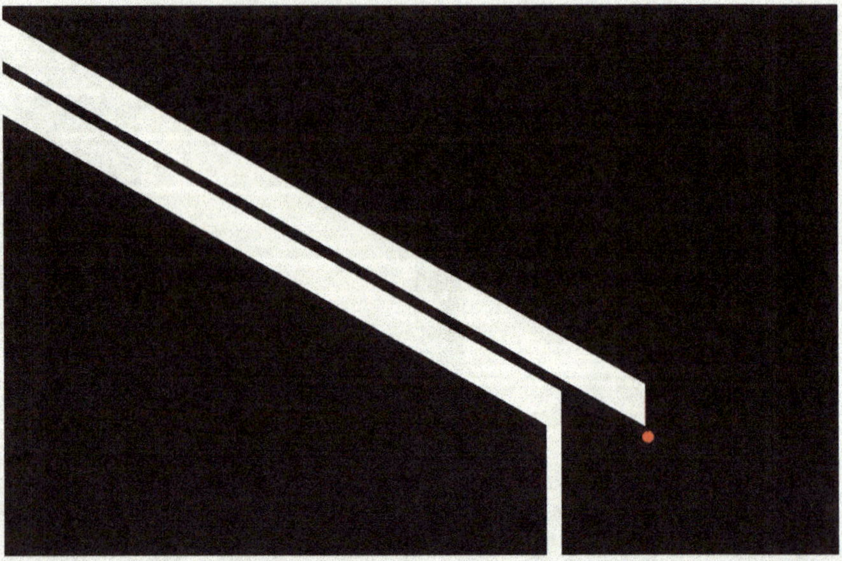

Because "the tool can do it" and "this tool can do your task" are separated by work that most people never do, and the assumption that they are the same is the most common reason a technically successful build produces nothing.

THE LINE

> ***If your AI is a shovel and you need a flathead, you are using the wrong AI.***

It is not a criticism of the shovel. Shovels are excellent. It is a statement about fit, and fit is a property of the pair, not of either item.

WHY THE GENERAL PURPOSE TOOL MAKES THIS HARDER

A large language model is a Swiss army knife. That is a genuine strength and it creates a specific trap.

With a shovel, you know quickly it is not a screwdriver. With a Swiss army knife, **there is always something that will engage with the screw.** You can get purchase. You can turn it a little. You will produce motion, and motion looks like progress.

So the failure is not refusal. **The failure is a mediocre result that is good enough to continue with and not good enough to rely on**, and because you got an output, you conclude the tool fits and you move on.

This is why chapter six insists on a standard written in advance. Without one, "it produced something" becomes the acceptance criterion by default, and every general purpose tool passes that test on every task.

THE THREE CONSTRAINT TYPES NOBODY CHECKS

Capability. Can it do this kind of thing at the quality your task needs. Not in a demonstration. On your inputs, including the ugly ones.

Permission. Systems are bound by rules, some deliberate, some emergent. Your task may sit somewhere the tool will not go, and you will discover this in week three, not in the sales call.

Context. The tool can do the thing in general and does not know your business. It does not know what your customers are like, what your margins are, which of your processes is held together with tape. That knowledge is what made your people good at the task, and none of it transfers by default.

Most fit failures are the third one, and the third one is the one that looks like a fit problem but is actually a discovery problem, which returns you to chapter two.

HOW TO TEST FIT HONESTLY

Give it the ten worst real examples, not ten typical ones. Typical cases pass. Automation earns or loses its money on the tail.

Do the test against the written standard, not against your impression. See chapter six, and see chapter eight for why your impression is not admissible.

Have the person who does the task now grade the output, blind if you can manage it. They are the only one who holds the standard, and by chapter seven they will not hold it forever, so use them while you have them.

Count how much you had to intervene. If getting a usable output took nine rounds of correction from an expert, you have not tested the automation. You have tested the expert.

THE HONEST CONCLUSION, WHICH NOBODY WANTS

Sometimes the answer is that no available tool fits, and the task stays manual.

That is a legitimate result and I deliver it regularly. **The businesses that take it well are the ones that had a written standard, because they can see exactly what the tool failed to meet.** The ones that take it badly are the ones evaluating against a feeling, and to them it sounds like an opinion.

BUILD AGAINST BUY, DECIDED HONESTLY

Fit failures produce a fork, and the fork is usually resolved by whoever is most tired rather than by analysis.

Buy when your task is genuinely common. If ten thousand businesses do this task roughly the way you do, somebody has built a tool for it, and that tool has absorbed edge cases you have not thought of yet. Your version will not be better and it will be yours to maintain forever.

Build when the thing that makes your task yours is the thing that makes it valuable. If your process is a competitive advantage, encoding it in a generic tool means encoding it as somebody else's assumptions.

And the tiebreaker that actually decides it: who will hold the standard in three years? A bought tool comes with a vendor's definition of correct, which is a real asset until your business diverges from it and nobody notices. A built tool has no definition of correct unless you wrote one down.

Either way the check is yours. **Buying moves the build. It does not move the check**, and the number of businesses that believe otherwise is the reason chapter nine exists.

FIVE QUESTIONS

1. For your last build, did you test on typical examples or on the worst ones?
2. How many rounds of expert intervention were needed to get a usable output?
3. Which of the three constraints, capability, permission, or context, has actually bitten you?
4. Who graded the output, and did they know what right looks like?
5. When did you last conclude that a task should stay manual, and how was that received?

12. WHEN THE TOOL DOES NOT EXIST YET

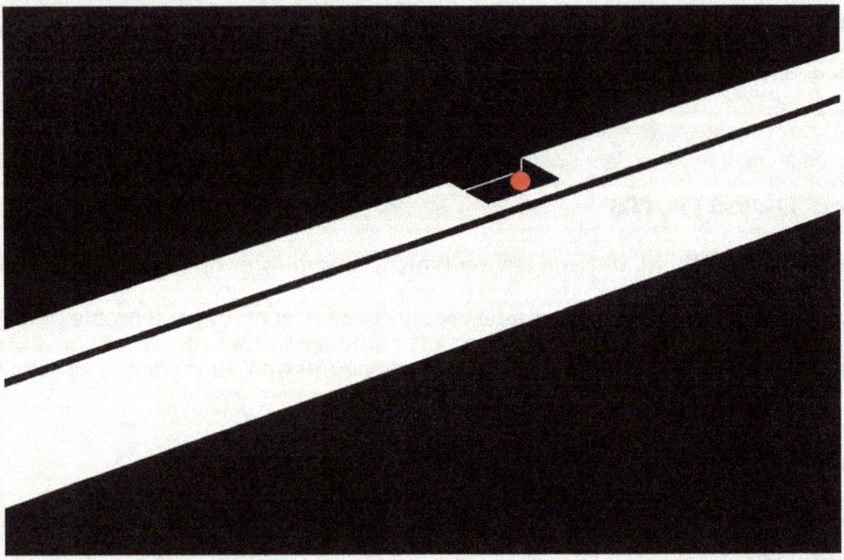

WHY THIS CHAPTER MATTERS

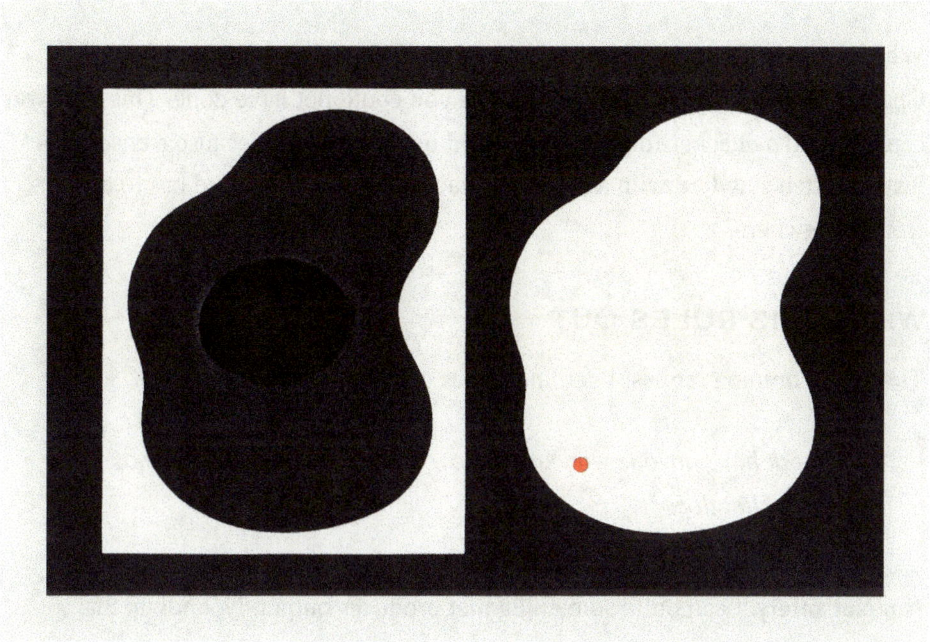

Because the honest answer to a fit problem is sometimes that you have to build the tool, and **building the tool has a prerequisite nobody states: you have to already be able to do the thing.**

THE RECURSION

I can train a system to do software design the way I do software design, and to fill gaps I might not think of, **because I know software design.** I can specify what good looks like. I can recognise a wrong answer. I can tell the difference between an unusual approach and a bad one.

If I did not know software design and asked a system to do software design, I would get an output. It would look like software design. **I would be working with a tool I was not equipped to refine, toward a standard I could not state, and I would have no way of knowing how far off I was.**

So:

> ***Building a specialised tool requires the expertise the tool is meant to supply.***

Which sounds like a paradox and is not. It is chapter five stated one level up. Chapter five says you cannot check work you could not have done. This says you cannot build a checker for work you could not have done, because a checker is just a written standard with a machine attached, and the standard has to come from somewhere.

WHAT THIS RULES OUT

The most common request I get, in various costumes:

> *"We do not have anyone who knows how to do this. Can you build us something that does it?"*

No. Not safely. I can build something that produces output. Nobody in the business will be able to tell whether the output is right, ever, and the arrangement will be unchecked from the first day. That is not an automation. It is a permanent unexamined dependency, and by chapter nine it will drift and nobody will notice.

What I can do, and what I say instead, is: **build the capability first, or buy access to someone who has it, and then automate.** The order is not negotiable. It is slower, more expensive, and the only version that survives contact with time.

THE LEGITIMATE VERSION

You have the expertise and you want to scale it. That works, and it is most of my good work.

You have a person who is excellent at the task. **You extract the standard from them**, which is chapter two done properly. You encode it. You test against the worst cases. **They grade the output.** You iterate until it meets the standard they hold.

Notice that the expert is present throughout, and notice that when you finish you must keep them in practice, or chapter seven takes the check away from you within a couple of years.

THE ONE THAT IS GENUINELY HARD

You have the expertise and it is one person, and that person is you, and you are the bottleneck you were trying to remove.

I do not have a clean answer. **What I have is a preference: automate the parts of your own work you could still do in your sleep, and keep doing the parts you are best at.** Not because the hard parts cannot be encoded. Because you are the only checker in the building, and if you stop doing the work, chapter seven comes for you specifically, and there is nobody behind you.

RENTING THE EXPERTISE, AND ITS LIMITS

There is a legitimate middle path and it gets abused, so it is worth stating the conditions.

You do not have the expertise in house. You bring in somebody who does, and they define the standard, specify the task, and grade the output until it holds. This works. Most of my better engagements are exactly this.

The failure comes at handover. The expert leaves, the automation runs, and the business now owns a system it could never have built and cannot evaluate. Day one of an unchecked dependency, purchased at consultancy rates.

So the deliverable cannot be the automation. It has to be the automation plus the standard plus a person who can apply it.

Written standard, in plain language, produced before the build and not reverse engineered from it afterwards. A named person inside the business trained to grade against it. And a scheduled point at which that person grades a real output while the expert is still reachable.

If a proposal does not include those three, it is not finished, whoever wrote it. I say that as somebody who sells these engagements, and it is the part clients push back on, because it looks like scope that does not produce anything.

It produces the only thing that keeps the rest of it working.

FIVE QUESTIONS

1. Is there anything you have automated in a domain where nobody in your business has the underlying expertise?
2. If yes, who is checking it, and against what?
3. For your best specialised automation, is the expert who defined the standard still practising?
4. What is the task you most want to remove from your own week, and are you the only person who can check it?
5. If you left for six months, which of your automations would still be trustworthy on your return?

PART IV

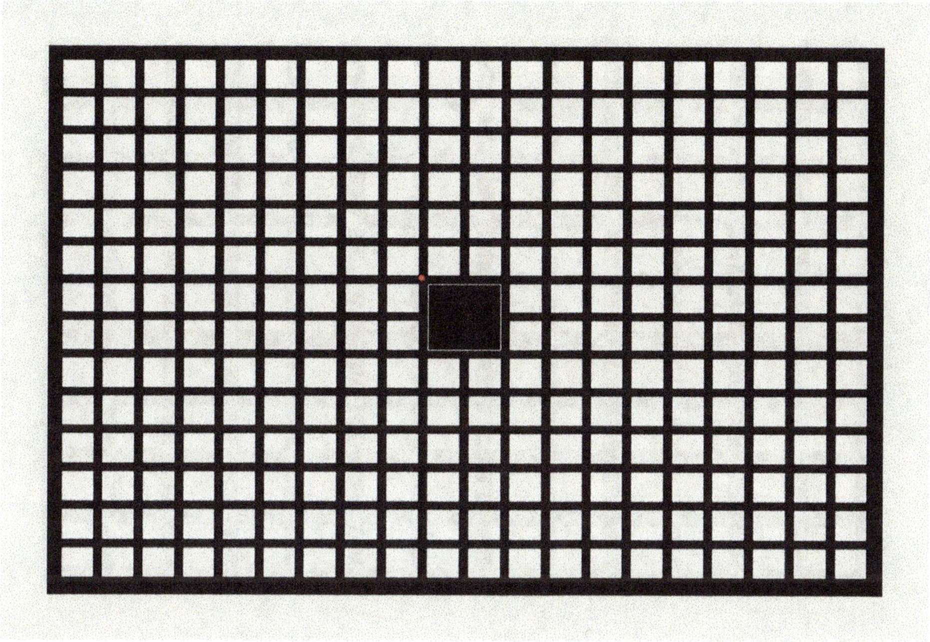

THE PEOPLE

13. THE WORK THAT STAYS HUMAN

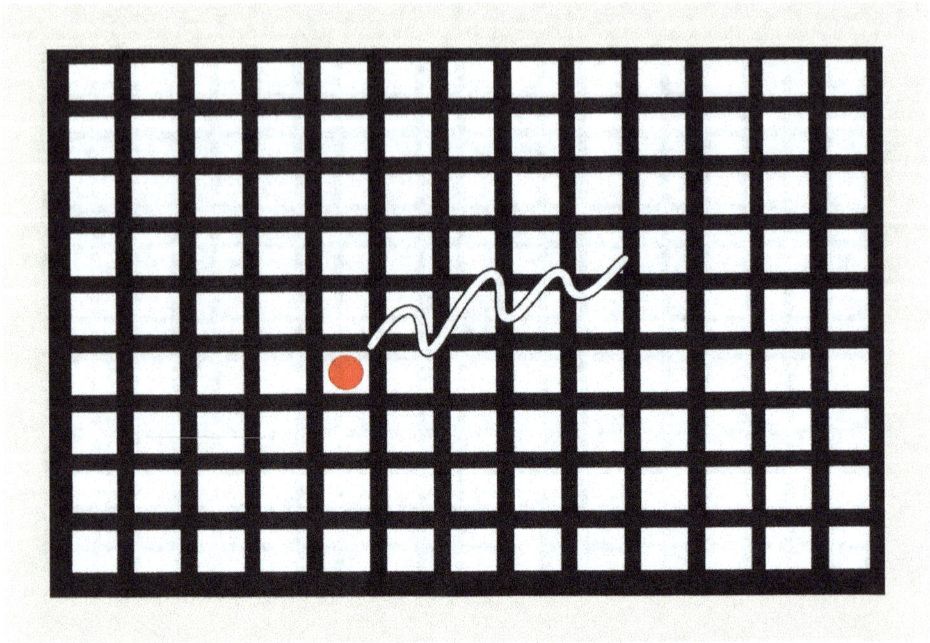

WHY THIS CHAPTER MATTERS

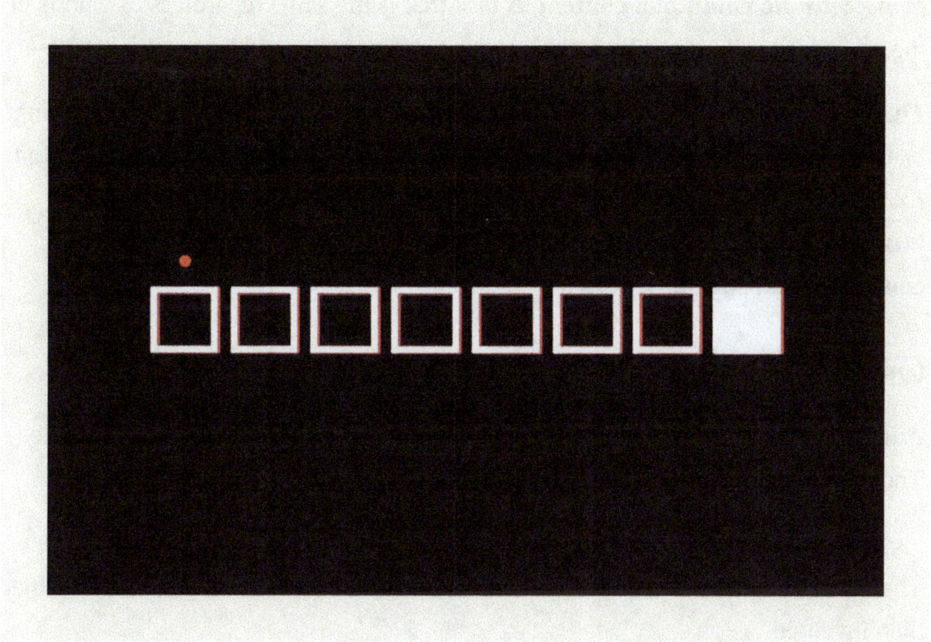

Because the gate in chapter three tells you what qualifies, and this chapter tells you what is excluded on principle, whatever the gate says. **There is a category of work I will not automate for a client, and I have never changed my mind about it in twenty years.**

THE RULE

> ***Where the work requires human judgment, human skill, human experience, or human emotion, it stays with a human.***

Not "stays with a human until the tools improve." Stays.

I want to be precise about the boundary, because this rule is often heard as a general anxiety about technology and it is not. It is narrow, and outside it I automate aggressively.

WHAT AUTOMATION IS GENUINELY GOOD AT

Work with no emotional content. A task that is the same regardless of who is on the other end of it, where the correct output does not depend on how anyone feels.

That is a large category and it includes most of what a business does. Scheduling. Data movement. Provisioning. Formatting. Routing. Reconciliation. Notification. Extraction. **These are not lesser work. They are the work that keeps a business running, and they are enormously improved by being done consistently at speed.**

WHERE IT BREAKS

Automation is black and white. It is direct, it is literal, and it does not modulate. Those are the properties that make it good at the category above, and they are disqualifying in a specific set of situations.

The example I use is deliberately extreme, and it is drawn from client work rather than invented.

I have worked with businesses whose customers are people dealing with traumatic events, and businesses working with people in therapy for mental health conditions.

You do not automate the contact. Not the intake, not the follow up, not the check in, not the difficult message. **Not because the automation would fail to produce appropriate text.** It would produce something appropriate. That is the problem. It would be appropriate and it would be empty, and the person receiving it is in a state where the difference between those two things is the entire content of the message.

The brokerage taught me this in a low stakes setting. Buyers stopped replying because they could feel the difference. **Move that same failure into a context where somebody is in crisis and the cost is not an open rate.**

THE GENERAL FORM

The narrow version of the rule is about emotion. The general version is about **whether the correct output depends on something the system cannot have.**

Judgment. Experience. Knowing this particular person. Having been there. Reading a situation that has not been described in the input, because nobody thought to describe it, or because it cannot be described.

That is the same thing that killed the brokerage sequence. The realtors knew something that was not in the record. In a follow up email that costs you a deal. **In a conversation with someone in a bad state it costs you something you cannot repay.**

WHAT I OWE YOU HERE, WHICH IS A LIMITATION

I have research behind most of this book. I do not have research behind this chapter, and I am not going to imply otherwise.

The claims in chapters five through nine are supported by controlled studies and a paper from 1983. **The claim in this chapter is twenty years of practice and no more than that.** I believe it firmly. I have not tested it, nobody I know of has tested it properly, and if you want to disagree with me you are disagreeing with my experience rather than with evidence.

I would rather say that plainly than dress an operating position up as a finding. This series exists partly to object to people doing exactly that.

THE QUALIFIER I KEEP USING, ON PURPOSE

I say "at least not yet" when I talk about this, and people notice, so let me be explicit.

I am not making a claim about what is permanently impossible. I do not know what these systems will be able to do. What I know is what they can do now, in the deployments I have seen, in businesses I am responsible for.

The rule is about the current state of the tool and the current cost of being wrong. If that changes, the rule should be re-examined. It has not changed yet, and I would want more than a demonstration before I moved it, because the failure mode here is not an inefficiency. It is a person receiving something hollow at the worst moment of their year.

AUTOMATE AROUND IT, NOT THROUGH IT

The rule does not mean these businesses cannot automate. It means the human contact is load bearing and everything else is fair game.

Automate the scheduling. Automate the record keeping, the reminders, the intake forms, the billing, the routing, the file preparation, the compliance logging. Automate everything that surrounds the conversation, and by doing so **buy the human more time to have it.**

That is the correct shape. The efficiency goes into the container, and the contents stay human. In practice it produces more human contact rather than less, which is the outcome everyone claims to want and very few configurations actually deliver.

FIVE QUESTIONS

1. In your operation, where does the correct output depend on knowing the specific person?
2. Have you automated any of those touch points?
3. If a customer received something technically appropriate and emotionally empty, what would it cost you?
4. What surrounds your human contact that could be automated to give people more time for it?
5. Are you automating the container or the contents?

14. WHERE IT WORKED, AND WHY IT WORKED

WHY THIS CHAPTER MATTERS

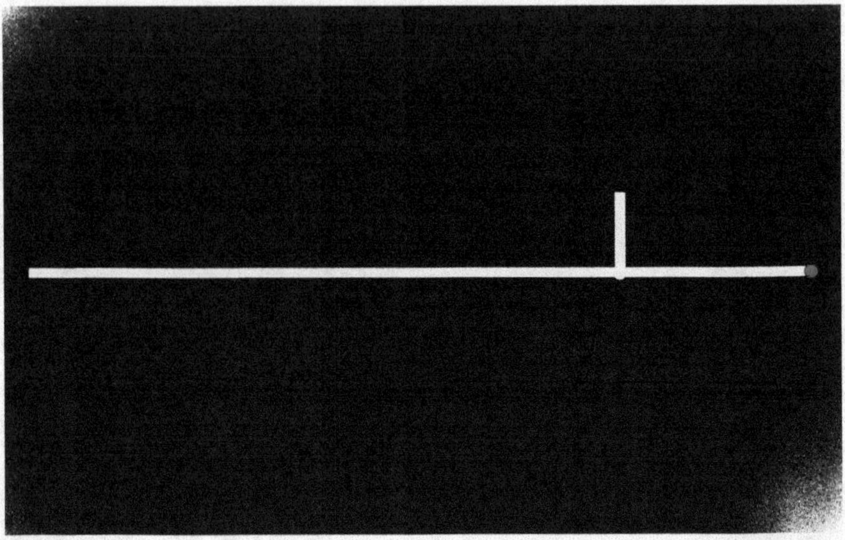

Because thirteen chapters of failure would give you a false picture. **I automate constantly and most of it works.** This chapter is about what the successful ones have in common, and it turns out to be the same finding stated positively.

THE ONE I DO MOST

Marketing and social media, across businesses from a thousand dollars a month to over a million a month, and inside companies you have heard of. It is the automation I recommend most often and the one that most reliably returns its cost.

And I want to be careful about why, because the reason is not that marketing is easy.

THE PRECONDITION

It works when the business already has:

- A defined brand voice
- A marketing plan and a business plan
- An ideal customer profile and an avatar
- A defined sales cycle
- Brand standards, fonts, colours
- Messaging aimed at a specific tier of a specific audience
- Analytics that are actually collected

When those exist, you can train a system on them. Voice, videos, blog posts, past writing, the standards document, the ICP, the sales cycle, the outcome you are aiming at. It has something to be trained against, and it can run with what the business already built.

When those do not exist, the same automation produces confident, well formatted, on brand looking output that is aimed at nobody. Same tool, same configuration, same operator. Different result, entirely because of what was in the building beforehand.

WHAT IS ACTUALLY GOING ON

Look at that list again. Brand voice, ICP, sales cycle, brand standards, analytics.

> ***Every item is a written standard.***

A business with a brand voice document has written down what correct sounds like. A business with an ICP has written down who correct is aimed at. A business collecting analytics has written down what correct achieves.

So the marketing automation works for exactly the reason the brokerage sequence failed. The brokerage had no written standard for the thing that mattered, so nobody could check it. A business with a brand manual has one, and everyone can.

This is not a chapter about marketing being a good automation candidate. **It is chapter six again, arrived at from the other direction.** The precondition for successful automation is that somebody already did the work of writing down what right looks like, and marketing is simply a function where businesses have historically been made to do that.

THE LOOP, WHICH IS THE PART PEOPLE SKIP

Setting it up is not the work. The work is the loop.

Refine, teach, and keep talking to it until the output is what you want. Then, and this is the step that gets dropped, **test it against the outcome you were actually after.** Not against whether you like it. Against the number.

Then feed the analytics back in. What performed, what did not. Now the system has evidence rather than instruction, and it will begin proposing shifts, and you test those too.

Done properly, it reaches a point where it anticipates what should come next. That is real and I have seen it repeatedly. **It is also earned over months and it is the direct product of the loop, not of the tool.**

THE SIDE EFFECT I CARE ABOUT MOST

Something happens to owners who run this loop properly, and it has nothing to do with marketing.

They start making decisions from data.

Marketing is the easiest place in a business to learn that habit, because the feedback is fast and the numbers are unambiguous. You publish, you see, you adjust. A pricing decision takes a quarter to evaluate. A campaign takes a week.

So the system ends up training the operator. It gets access to the analytics, it reports what is working before the owner would have looked, and it proposes changes with reasons attached. **The owner learns what evidence based decision making feels like on a fast loop, and then applies it elsewhere.**

I did not design that outcome and I now consider it the most valuable thing the engagement produces.

AND THE LIMIT, STATED IN THE SAME BREATH

> ***These systems are very good at using data to make decisions. They are not built to understand the emotional side of your business.***

For this task, that is not a weakness. It is the specification. **You do not want your marketing automation making emotional decisions**, and the fact that it cannot is precisely why it is safe here and unsafe in chapter thirteen.

The tool did not change between those two chapters. **The task did.**

WHY THIS IS NOT A COUNTEREXAMPLE TO THE BOOK

It would be easy to read chapter one against chapter fourteen and conclude that email automation failed and marketing automation worked, so it depends.

That is not the finding. Both are marketing. Both are automated messages to prospects. **One had a written standard for what made the message work and one did not**, and that alone predicted the outcome.

The presence of the check is the variable. Everything else is detail.

FIVE QUESTIONS

1. Which of the preconditions in this chapter does your business actually have in writing?
2. For your best performing automation, what written standard existed before it?
3. Are you running the loop, or did you set it up and leave it?
4. When did you last feed real outcome data back into something you automated?
5. What has automation taught you about how you make decisions?

15. THE ROLES IT TAKES

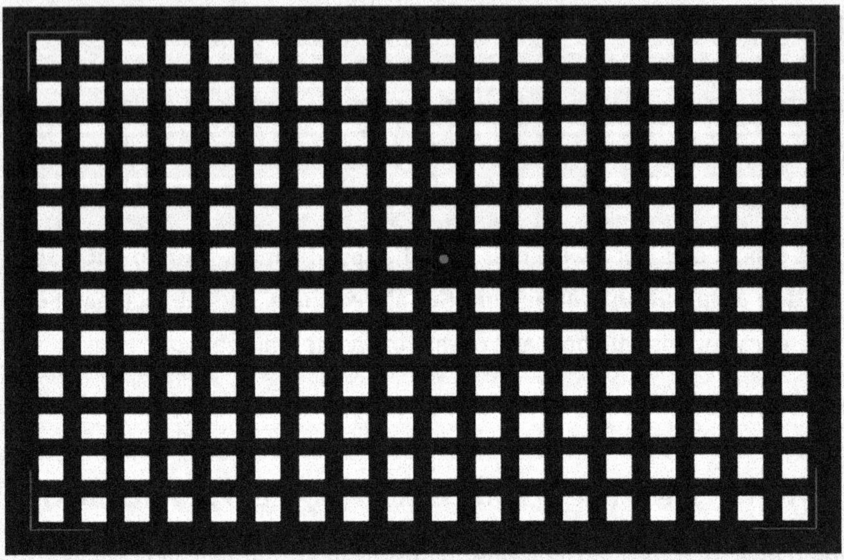

WHY THIS CHAPTER MATTERS

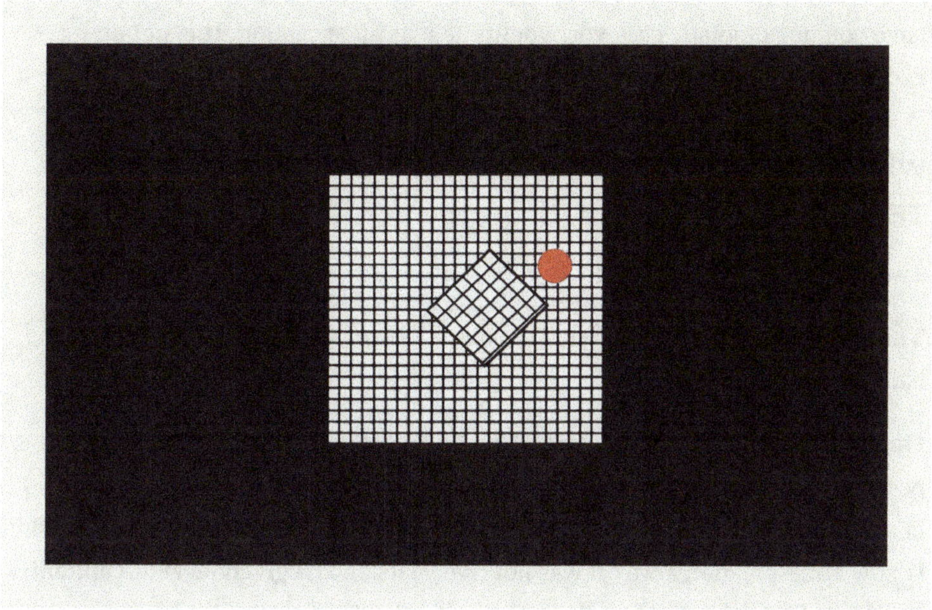

Because everything before it has been about whether an automation works, and this is about what it costs someone when it does. **I have put in automations that ended people's roles. I am not going to write around that.**

THERE IS NO CASE STUDY IN THIS CHAPTER, AND THAT IS THE POINT

Every other chapter in this book has a story. This one does not, and I thought hard about whether to invent a representative one. I am not going to, because the truth is more uncomfortable than a single story would be.

It is no longer one engagement. It is the direction of the work.

Email automation. Marketing automation. Customer service automation. Automated phone handling. Chat. Nurturing sequences. A large and growing share of what I am hired to build will either end somebody's role or change it into something they did not sign up for.

That is not a prediction. It is a description of my current practice, and it has become true gradually enough that I did not have a single moment of noticing. If I gave you one dramatic case you would file it as an exception. **It is not an exception. It is the arithmetic of the job.**

WHY IT IS ACCELERATING, AND IT IS NOT ONLY THE TECHNOLOGY

Two things are compounding.

The systems are getting better. That one is obvious and it is not the interesting half.

Operators and integrators are getting better at using them. That is the half people miss. The constraint on automation has rarely been raw capability. It has been the skill of the person specifying, scoping, testing and installing it. That skill is now widespread in a way it was not five years ago. **A given level of capability produces far more displacement in the hands of a practised integrator**, and there are far more practised integrators than there were.

I am one of them. That sentence is not comfortable to write and it is the honest position from which this chapter is written.

THE CLAIM I HAVE TO BOUND CAREFULLY

Here is what I believe, and here is the boundary on it, because the unbounded version is wrong and I would have written the unbounded version a year ago.

Inside the gate, machine error is lower than human error. A task that is genuinely repeatable, needs no customisation, and takes no thought is precisely the kind of task machines have beaten people at since the assembly line. That is not a claim about AI. It is the whole history of industrial automation. On an automotive line running the same operation ten thousand times, the machine is more accurate than the person and it is not close, and pretending otherwise helps nobody.

Outside the gate, the evidence runs the other way. Chapter eight has the studies. People given an AI assistant wrote less secure code than people without one. Experienced developers on familiar codebases were measurably slower. **The moment a task requires judgment or customisation, it has left the gate, and the error advantage inverts.**

So the three variables are not only a quality filter. **They are the line where the error claim changes sign.** Inside, the machine wins on accuracy. Outside, it does not, and the confident belief that it does is itself the failure.

This matters enormously for this chapter, because "the automation is more accurate than the person" is the sentence used to justify ending a role, and it is only true on one side of a boundary that almost nobody draws.

THE RULE I CARRIED FROM VOLUME IV, AND IT APPLIES ABSOLUTELY

In Volume IV, people lost their jobs at a restaurant because nobody had watched margin for years. I wrote then that **the layoff was the cost of a measurement failure and never a strategy**, and that framing is not softening. It is the more damning reading.

The same rule governs here, and it is stricter.

If you end a role because the task genuinely passed the gate and the machine does it better, that is a real operational decision and you should own it as one. **If you end a role because you assumed the machine does it better, and the task never passed the gate, you have not made an efficiency. You have made an error and charged it to a person.**

The second one is far more common than the first. It is more common because the assessment in chapter four is almost never run at task level, so the decision gets made at job level, where the forty percent that was never eligible is invisible.

I have watched businesses eliminate a role that was seventy percent automatable and rehire eighteen months later for the thirty percent, at a higher salary, from a worse pool. Everyone involved described the original decision as an efficiency. It was a costly mistake with a person's livelihood inside it.

THE ALTERED ROLE, WHICH NOBODY COUNTS

The endorsement this book sits under says you are accountable for the roles automation **displaced or altered**. The second word does more work than the first.

When somebody stops doing the work and starts checking the machine's work, we say the role was preserved. On the org chart it was.

By chapter seven, it was not. They are now doing the residue, the part nobody could work out how to encode, stripped of the routine work that used to make the day coherent. They are monitoring, which is the thing humans are worst at. Their skills are decaying because they no longer practise. And they are the person expected to take over when it fails, which is by definition the hardest version of the task.

Nobody was displaced. Nobody appears in any statistic. **And the job got worse in four distinct ways, none of which were anybody's stated intention.**

If you do this to someone, at minimum know that you did it, and staff it the way Bainbridge said to: **more skilled than average, less loaded than average.** I have never seen a business do it. I had never proposed it either, until this book was researched.

THE NUMBER I AM NOT GOING TO LET YOU USE

You will have heard that forty seven percent of jobs are at risk of automation. It is probably the most quoted figure in this entire subject.

It comes from a 2013 Oxford paper, and the paper is careful and honest. **The quotation of it is neither.**

Here is what the paper actually says. "**About 47 percent of total US employment is at risk.**" At risk. Occupations rated susceptible, not jobs forecast to disappear.

Here is how the number was produced. The authors, with a group of machine learning researchers, **subjectively hand labelled seventy occupations** at a workshop, marking each one automatable or not, and that labelling was extended across more than seven hundred occupations.

Here is the timeframe. **"Some unspecified number of years, perhaps a decade or two."**

And here is the authors' own limit, which nobody quotes: they focus on the mix of jobs that existed in 2010, they restrict themselves to the substitution effect, and they **make no attempt to forecast future occupational composition.**

In 2019 a pair of economists tested the predictions against what actually happened to US employment between 2013 and 2018. Their finding was that compared with standard approaches based on routine task intensity, **the predictions did not add value for forecasting the impact of technology on employment.**

That is the fifth widely quoted number this series has taken apart, after four about AI project failure in Volume V. **The pattern is identical every time: a careful paper, a careless quotation, and a decade of people making decisions on a sentence the authors did not write.**

I am not telling you automation does not cost jobs. **I have just spent this chapter telling you that it does, in my own practice, repeatedly.** I am telling you that the number in your head is not evidence, and that if you are going to end someone's role you owe them a better basis than a statistic that failed its own backtest.

WHAT I THINK IS HAPPENING, OFFERED AS OPINION

I do not think this is disruption in the way the word is being used.

I think it looks like the internet. People called it a fad, then a revolution, then it quietly ended up inside every part of every business and stopped being a topic. **I expect the same shape here: less dramatic than the loudest voices claim, more total than the sceptics expect, and slower than both.**

That is a prediction and I want it marked as one. I have no evidence for it. It is a pattern match against one previous technology, made by somebody with an obvious stake in the outcome. **Nothing in the research behind this book settles it, and you should weight it accordingly.**

THE PART I HAVE NOT RESOLVED

I am going to end without tidying this up, because tidying it would be dishonest.

I hold two things at once and they do not reconcile.

The first is that when a task genuinely passes the gate, automating it is the right call. It is more accurate, it is cheaper, it frees people for work that needs them, and refusing to do it out of sentiment is not kindness. It is a slower failure, and in a small business a slower failure still takes everybody's job eventually.

The second is that a person had a role and now does not, and I built the thing that did it. Sometimes I have known their name.

I do not think the first cancels the second. People who do this work often talk as though it does, and the ones who talk that way most confidently are usually furthest from the room where it was announced.

So the honest position, and the one I will leave you with, is this. **The efficiency is real and the cost is real and they do not net off.** You carry both. What you owe the person is that the decision was made properly, at task level, against the gate, with the error claim bounded where it should be bounded, and not on the basis of a number somebody read in a headline.

That is not absolution. It is the difference between a hard decision and a careless one, and it is the only part of this that is in your control.

FIVE QUESTIONS

1. Has an automation of yours ended or materially changed someone's role?
2. Was that decision made at task level or at job level?
3. Did the task actually pass the gate, or was it assumed to?

4. For roles that were altered rather than ended, is that person more skilled and less loaded than average, or the opposite?
5. What basis did you give the person for the decision, and would it survive being read back to you?

BACK MATTER

THE AUTOMATION OPERATOR'S DIAGNOSTIC, RETAKEN

	BEFORE	NOW	IN 90 DAYS
The task, written end to end			
The standard, wrong output described			
The checker, named and able today			
The decay, still able in three years			

Part three is the one to keep. It is four questions long, you can run it on any task in your business this afternoon, and it is the only part that answers the question the rest of the book is about. The other three describe the work. Part three names the person who could still do it, and if that name does not exist the automation is not being checked by anybody, whatever the dashboard says.

THE CLOSING QUESTION

> ***If it were wrong tomorrow, who in your building could still tell?***
>
> ***If you cannot name them, it is not automated. It is unattended.***

A NOTE ON METHOD

WHERE THIS BOOK CAME FROM

The argument is drawn from an interview conducted on 17 August 2026, covering eleven questions and fifteen follow ups. **The anchor case is a single client engagement: a twenty agent real estate brokerage, 2024.** One case, stated as one case. Nothing general about email automation is claimed from it.

HOW SOURCES WERE HANDLED

Every source was placed in one of four tiers. **Tier A** means the primary text was located and read. **Tier B** means located and not yet read, and nothing at Tier B is printed. **Tier C** means contested or vendor sourced, and is printed only with the weakness stated. **Tier D** means no adequate source, and is either cut or stated as experience and labelled as such.

Three Tier A sources carry the book.

Bainbridge, L. "Ironies of Automation." ***Automatica*****, Vol. 19, No. 6, pp. 775 to 779, 1983.** Chapter seven is built on it, and the fourth condition of the instrument is hers rather than mine. **She is writing about process control, not knowledge work.** The transfer is argued in chapter seven and is not assumed. I had not read this paper before this volume was researched, and chapter seven says so in its second paragraph.

Perry, N., Srivastava, M., Kumar, D., and Boneh, D. "Do Users Write More Insecure Code with AI Assistants?" Proceedings of the 2023 ACM SIGSAC Conference on Computer and Communications Security. The model studied is now several generations old. **The finding used here is about the human relationship to the tool, not about that model's capability**, and no claim about any current system is made from it.

"Measuring the Impact of Early-2025 AI on Experienced Open-Source Developer Productivity." METR, 10 July 2025. Sixteen developers, 246 issues. The authors state explicitly that they do not claim their developers or repositories represent most software development work, and that limit is printed in chapter eight alongside the finding.

Frey, C. B., and Osborne, M. A. "The Future of Employment: How Susceptible Are Jobs to Computerisation?" Oxford Martin School, 2013, and **Coelli, M. B., and Borland, J. "Behind the Headline Number." Melbourne Institute Working Paper 10/19, 2019.** Both read. Chapter fifteen.

FIVE NUMBERS THIS SERIES WILL NOT PRINT

Four statistics about AI project failure were taken apart in Volume V: a misquoted Gartner prediction about bias, an unsourced vendor figure, a misattributed RAND number, and an unreviewed pilot failure rate. **The forty seven percent employment figure is cut in chapter fifteen.**

None of the five is used anywhere in this book, including as an illustration of what not to use, except in the passages where it is dismantled.

WHAT IS CLAIMED AS EXPERIENCE AND NOT AS EVIDENCE

Chapter thirteen has no research behind it and says so in the chapter. The rule about work that stays human is twenty years of practice. It is stated as a position, not a finding.

The rate at which I find errors in generated code is my impression of my own work. No percentage appears anywhere in this book. I do not count, and a figure I did not measure is exactly the kind of thing this series exists to object to.

The claim that automation increasingly costs roles is my description of my own practice. No industry figure is offered, because the available ones do not survive checking.

The comparison with the internet in chapter fifteen is a prediction, is marked as one, and has no evidence behind it.

ON PRODUCTS

No AI vendor, model or company is named in the body of this book. The text says "current large language models." Named models appear in this note only, because a citation that hides its object cannot be checked by a reader, and a book that asks you to verify its sources has to make that possible.

ON THE INSTRUCTION THIS VOLUME WAS WRITTEN UNDER

The author's instruction when this volume was commissioned was to be methodical, keep it cohesive, and **not let it become a record of his own biases.**

Two passages exist because of that instruction. **Chapter eight contains a study that cuts against the author's own position** on expertise, and it is presented as a correction rather than buried. **Chapter fifteen bounds a claim the author made in stronger terms**, because the unbounded version contradicted both his own account and the available evidence.

Where his experience is unsupported, this book says so in the body rather than in a footnote.

REFERENCES

Bainbridge, Lisanne. "Ironies of Automation." *Automatica* 19, no. 6 (1983): 775 to 779. *Written about process control rather than knowledge work. The transfer is argued in chapter seven and is not assumed.*

Coelli, Michael B., and Jeff Borland. "Behind the Headline Number: Why not to Rely on Frey and Osborne's Predictions of Potential Job Loss from Automation." Melbourne Institute Working Paper 10/19, 2019.

Frey, Carl Benedikt, and Michael A. Osborne. "The Future of Employment: How Susceptible Are Jobs to Computerisation?" Oxford Martin School, 2013. Published in *Technological Forecasting and Social Change*, 2017. *Seventy occupations were hand labelled subjectively and extrapolated across 702. The paper claims susceptibility, not outcome, and says so. Chapter fifteen quotes it on that point.*

METR. *Measuring the Impact of Early-2025 AI on Experienced Open-Source Developer Productivity.* 10 July 2025. *Sixteen developers, 246 issues. The authors state that they do not claim their developers or repositories represent a majority or plurality of software development work. That limit is printed in chapter eight alongside the finding.*

Perry, Neil, Megha Srivastava, Deepak Kumar, and Dan Boneh. "Do Users Write More Insecure Code with AI Assistants?" *Proceedings of the 2023 ACM SIGSAC Conference on Computer and Communications Security.* Association for Computing Machinery, 2023. *The model studied is several generations old. The finding used here is about the human relationship to the tool, not about that model's capability.*

ABOUT THE AUTHOR

JT Taylor, MS, MBA is an operations executive and founder. He holds a Master of Science and an MBA in entrepreneurship and marketing innovation.

He integrates scalable systems, software and operations for business owners globally, and has worked with organizations from pre-launch startups to billion dollar companies across service, home services, hospitality, medical, wellness, fitness, consumer goods, real estate, law, ecommerce, construction and education.

He came to psychology on the way to surgery. While weighing a path toward becoming a doctor of psychology, he spent a year studying army psychology, entirely by choice and not as part of any requirement. He concluded he did not want to enter the army and never served. What stayed with him was the subject itself, and the conviction that how the person at the top of a business thinks determines almost everything that happens beneath them. He profiles the owner before he examines the business.

He is the founder of the National Board of Operations Professionals, and hosts a podcast on operator-class business insights.

Principles of Practice is a series. This is Volume VI, the final volume.

www.ingramcontent.com/pod-product-compliance
Lightning Source LLC
LaVergne TN
LVHW010736061126
851857LV00014B/1512
* 9 7 9 8 9 5 2 4 6 6 0 5 0 *